1

This textbook is designed to fulfill the mathematics curriculum requirements for community elementary schools. It is being proposed and recommended as a mathematics text especially for use by pupils in the fifth grade in English-speaking countries.

It contains valuable information to help pupils understand fundamental principles and concepts, integrated with numerous problems with answers, of Elementary Mathematics. Other titles in J. Nyenetu Jarkloh's series of Elementary Mathematics are:

Modern Mathematics for Community Elementary Schools (Grade 5): Teacher's Edition

Principles of Modern Elementary Mathematics (Grade 6)

Principles of Modern Elementary Mathematics (Grade 6): Teacher's Edition

MODERN MATHEMATICS FOR COMMUNITY ELEMENTARY SCHOOLS

GRADE 5

J. NYENETU JARKLOH

DEDICATION

This book is dedicated:
To the loving memory of my dear mother,
Madam Jlopleh Jugbe;
To the loving memory of my beloved son,
Karpeh Jarkloh;
And to
Dr. James Teah Tarpeh,
who had given me so much courage and inspiration as well as moral and
material support to continue to move on during the darkest period in
Liberia's recent past, I dedicate this book.

Table of Contents

Dedication ------------------------------------ 5
Acknowledgements ------------------------------------ 11
Preface --- 13

Chapter One: Operations with Numbers ------------------ 17
1.1 An Introduction to Arithmetic Concepts and Operations 17
1.2 The Order of Arithmetic Operations ---------------------- 21
1.3 Signs of Divisibility ------------------------------------ 22
1.4 The Natural Number Series ------------------------ 24
1.5 The Concept of Place Value in the Writing
    and Reading of Numbers ------------------------------- 25
1.6 Roman Numerals and Number Bases ---------------------- 28
1.6.1 Operations with Number Bases ---------------------- 29
1.7 Rounding off of Natural Numbers ---------------------- 32
1.8 An Introduction to Geometric Concepts --------------- 34
1.9 The Number Line. Comparison of Natural Numbers ---- 39
1.10 Rational Numbers ------------------------------- 41
1.11 Coordinate Straight Line ---------------------------- 43
1.12 The Module of a Number ---------------------------- 45
1.13 Comparison of Rational Numbers --------------------- 47
1.14 Rectangular System of Coordinates ----------------- 48
Chapter One: Questions to Test Your Understanding ------ 52
Chapter One: Problems and Exercises --------------------- 55
Chapter One: Exercises to Rate Your Ability ---------------- 63

Chapter Two: Addition and Subtraction ---------------- 66
2.1 Properties of Addition ------------------------------- 66
2.2 Properties of Subtraction --------------------------- 67
2.3 Addition and Subtraction Tables ----------------------- 70
2.4 Oral Addition and Subtraction Operations ------------ 72
2.5 Written Addition and Subtraction Operations ----------- 73
2.5.1 Written Addition Operations with Natural Numbers --- 73
2.5.2 Written Subtraction Operations with Natural Numbers -- 73
2.5.3 Addition of Rational Numbers ------------------------ 74

2.5.4 Subtraction of Rational Numbers   ----------------------- 77
  2.5.5 More on Powers or Exponents   ------------------------ 80
2.5.6 Coefficients   --- 83
2.5.7 Basic Properties of Equation   -------------------------- 84
2.6 Angles and Polygons   ------------------------------------- 86
2.6.1 Angles   --- 86
2.6.2 Polygons   --- 90
2.6.3 Classification of Triangles   ---------------------------- 91
2.7 Physical Quantities   ------------------------------------- 94
2.8 Expressions, Equations, and Inequalities   ---------------- 98
2.8.1 Equations and Equalities   ------------------------------ 98
2.8.2 Inequalities   --- 100
2.9 Scale   -- 101
2.10 Reading Problems   -------------------------------------- 102
Chapter two: Questions to Test Your Understanding   ----- 107
Chapter Two: Problems and Exercises   -------------------- 108
Chapter Two: Exercises to Rate Your Ability   --------- 121

Chapter Three: Multiplication and Division   ---------- 123
3.1 An Overview of the Concept   --------------------------- 123
3.2 Multiplication and Division of Natural Numbers   ---- 125
3.3 Commutative Property of Multiplication   -------------- 126
3.4 Associative Property of Multiplication   ---------------- 128
3.5 Distributive Property of Multiplication and Division ------ 129
3.6 Written Multiplication Operation with Natural Numbers - 130
3.7 Written Division Operation with Natural Numbers   --- 132
3.8 Multiplication of Rational Numbers ---------------------- 135
3.9 Division of Rational Numbers   -------------------------- 137
3.10 Circle and Circumference   ----------------------------- 139
3.11 Areas of Rectangle and Triangle   ---------------------- 141
3.12 Rectangular Parallelepiped   --------------------------- 144
3.13 Volume of a Parallelepiped   --------------------------- 145
Chapter Three: Questions to Test Your Understanding   ---- 147
Chapter Three: Problems and Exercises   -------------------- 148

Chapter Four: Common Fractions   ----------------- 160
4.1 Fractions and Fractional Numbers   --------------------- 160

4.2 Proper and Improper Fractions ------------------------- 163
4.3 Comparison of Common Fractions --------------------- 166
4.4 Basic Property of Fractions ------------------------- 168
4.5 Reduction of Fractions to a Common Denominator ------ 170
4.6 Addition and Subtraction of Fraction
      with Identical Denominators ------------------------- 171
4.7 Addition and Subtraction of Fractions
      with Unlike Denominators ------------------------- 174
4.8 Conversion of Common Fractions to Decimals ------- 175
4.9 Decimal Approximations of Common Fractions ------ 176
4.10 More about Common Fractions ---------------------- 179
4.10.1 Mutually Inverse Numbers ---------------------------- 179
4.10.2 Multiplication of Common Fractions --------------- 179
4.10.3 Division of Common Fractions --------------------- 182
4.10.4 Finding of a Number by its Fraction ---------------- 183
Chapter Four: Questions to Test Your Understanding -------- 184
Chapter Four: Problems and Exercises ------------ 185
Chapter Four: Exercises to Rate Your Ability -------------- 194

Chapter Five: Decimal Fractions ------------------ 196
5.1 The Concept of Decimal Fractions --------------------- 196
5.2 Properties of Decimal Fractions ------------------------ 199
5.3 Comparison of Decimal Fractions --------------------- 200
5.4 Rounding off of Decimal Fractions ------------------ 201
5.5 Addition and Subtraction of Decimal Fractions ----- 202
5.6 Multiplication of Decimal Fractions ------------------ 205
5.7 Multiplication and Division of Decimal
      Fractions by 10, 100, and 1000 ---------------------- 208
5.8 Division by Decimal Fractions ---------------------- 212
5.9 Arithmetic Mean -- 214
Chapter Five: Questions to Test Your Understanding ------ 215
Chapter Five: Problems and Exercises -------------------- 216

Answers to Problems and Exercises ----------------- 228

Appendix --- 234
I. Tables of Weights and Measures -------------------- 234

The Metric System ----------------------------------- 234
The British (or Imperial) System --------------------- 235
II. Useful Equivalents ----------------------------------- 235
III. List of Abbreviations and Symbols ------------------ 236
IV. The Greek Alphabet ------------------------------- 236
V. A Table of Common Polygons and Polyhedrons
and Their Formulae ----------------------------------- 237

ACKNOWLEDGEMENTS

The writing of a book is a tedious and complex undertaking. I would like to express my thanks to all those who, during the preparation of this book, have contributed to it in no small measure. In particular, I acknowledge the contributions of Dmitry Ilich and Alexandr Igorivich of Orion Group of companies in drawing some of the geometric figures.

I acknowledge the contribution of Dr. Vyacheslav S. Shobik, Associate Professor and Chief of Foreign Affairs Department of Odessa National Polytechnic University. Dr. Shobik had reviewed and pointed out some mistakes contained in the original text of the book; and, despite his busy time at work, he was always available when I needed to consult with him. I am sincerely grateful for his editorial and moral support.

I am also very grateful to Evgeniy N. Bogdanov, owner of the Sevastopol-based company called "BrainCrackers.com". He was helpful in converting the MS Word interior file of the book into the PDF format required for publication.

My special thanks go to my son Julian Jarkloh for also drawing some of the geometric figures, suggesting and doing the design of the front cover.

My friend Nikolai Arminevich Danilchenko of TransSibGas accorded me the hospitality to print out the finished text of the book in preparation for publication. I am thankful for his usual material assistance, which was most valuable in getting the book ready for publication.

Finally, this book would never have been completed without the moral support of my wife Ella, who always believed in me even in those difficult circumstances when I felt that it wasn't possible. I am sincerely grateful for her support, patience and encouragement while I wrote this book.

PREFACE

All initial manuscripts I had prepared were lost in the wake of the Liberian Civil War. These included "Principles of Electronics" (materials of my lectures in electronics at the W.V.S. Tubman College of Technology in Harper, Maryland County), "Chemistry Made Easy for Liberian High Schools" (materials of my lectures in chemistry at the Cape Palmas High School), "Elements of College Mathematics" (materials of my lectures in Numerical Analysis at the University of Liberia), "Anthology of a Country Boy", and "A Big Heart and a Small Fortune". The "Anthology of a Country Boy" was a collection of poetry manuscripts I had written in my high-school and college days. Some of these were published in the New Liberian Newspaper where I worked as a reporter and editorial assistant while studying at the University of Liberia (UL). "A Big Heart and a Small Fortune" was another collection of poems I had written during my studies at the UL and in the former Soviet Union. I revised and updated these manuscripts after my return home at the William V. S. Tubman College of Technology where I took up teaching assignment as an instructor up to the inception of the Liberian civil crises. It is a great pity when I think of the tremendous efforts I had invested in said manuscripts. I, therefore, feel sincere joy that this present book you are holding, "Modern Mathematics for Community Elementary Schools: Grade 5", has finally become a reality.

"Modern Mathematics for Community Elementary Schools (Grade 5)" is an attempt to make a modest contribution by preparing a comprehensive mathematics text that would meet the needs of fifth grade pupils and their teachers as well as respond to the elementary-level mathematics curriculum development objective of the Liberian educational system in particular and the West African regional educational system in general. The quality of the text fulfills international standard and, therefore, the book can be used in any part of the world. The materials and concepts presented in it are designed and intended to give a solid foundation in mathematics mainly to fifth grade students. While it has been prepared primarily for pupils, it is also hoped that fifth grade mathematics teachers would find the text useful in the planning and presentation of their lesson plans. In this direction, the workbook in the Teacher's Edition would prove very helpful to teachers.

The materials featured in the textbook represent many years of teaching experience in various local elementary and junior high schools in Monrovia, as well as in the Johnny Vorker Junior & Senior High School in Saclepea (Nimba County, Liberia) and the Kpadeh Gissi Elementary School on the Block Path near Firestone Plantations Company Division 25. Of course, considerable part of the materials is based on extensive and personal research efforts to make it as compatible as possible with international standard.

For many years it has been my desire to write such a text based on indigenous experiential background. My desire stems from the absence of textbooks written by West African scholars and the simple fact that the presence of textbooks written by African scholars would make a tremendous difference in awakening the academic interests of pupils in study materials presented in the context of their cultural and traditional experience. This, I believe, would produce a number of positive effects relative to the enhancement of academic performance and an overall educational result.

"Modern Mathematics for Community Elementary Schools (Grade 5)", divided into 5 chapters, is the first of my two books dealing with mathematics on the elementary level. Some of the chapters are followed by questions which test pupils' understanding of basic concepts dealt with. Besides, ample problems and exercises are given at the end of each of the five chapters. Answers to almost all of these problems and exercises are provided at the end of the book. It is recommended that pupils attempt to solve said problems and exercises and compare their answers either to those provided at the end of the textbook, or to those in the practice book "Solutions to Problems and Exercises: A Workbook (Grade 5)". This workbook contains solutions with explanations to almost all of the *Problems and Exercises* at the end of each chapter of this book.

Additional exercises are given at the end of some chapters which are intended to rate pupils' ability. People learn by doing; therefore, in order to strengthen their knowledge and understanding, pupils are encouraged to make every effort to answer the questions and solve the problems at the end of every chapter.

Finally, I sincerely invite critical comments and suggestions that could help improve "Modern Mathematics for Community Elementary Schools (Grade 5)" in the context of the curriculum development objective of the country-specific educational system in particular and the West African regional educational system in general.

J. Nyenetu Jarkloh.

CHAPTER ONE: OPERATIONS WITH NUMBERS

1.1 An Introduction to Arithmetic Operations and Concepts

Definitions

Arithmetic is a branch of mathematics concerned with numerical calculations such as addition, subtraction, multiplication, and division. You are already acquainted with the aforementioned arithmetic operations and concepts from the first grade, through the second, third, and up to the fourth grade. That is why this present topic cannot be considered an introduction, but rather a repetition of definitions just to refresh your memory about what you already had studied in your mathematics class in previous years. There is no need for you to worry, however, if you do not already possess the necessary background in mathematics. This present textbook has been designed with the aim to provide you with fundamental theoretical concepts and explanations, ample examples, exercises, and problems based on practical considerations. Answers and solutions to problems and exercises, which are intrinsic feature and platform by which the presentation of theoretical concepts is driven, would ensure that you acquire the level of numerical and operational ability required to succeed, if not excel, in your study of mathematics beginning at this point. For this reason, it is recommended to make concomitant use of the workbook containing solutions to the problems in this book.

Before going further, it is essential to clarify the four basic arithmetic operations.

The concept about what *addition* is arises from such simple fact that it does not require any definition and may not be formally defined. Numbers, which are added together, are called *addends*. A number that is gotten as a result of addition is called *sum*. For example, 8 + 6 = 14. In this operation, 8 and 6 are the addends, and 14 is their sum. The sign plus (+) is used to designate an addition operation.

Subtraction is the operation of finding the *difference* between two numbers by means of decreasing one of the numbers (*minuend*) by the quantity of the other number (*subtrahend*). For example, 12 − 5 = 7. 12 is the minuend; 5 is the subtrahend; and 7 is the answer or difference (sometimes called the *remainder*). The sign minus (−) is used to designate a subtraction operation.

Multiplication is an operation, defined initially in terms of repeated addition, by which the result of two quantities or numbers is calculated. To multiply a number (*multiplicand*) by a whole number (*multiplier*) means to repeat the multiplicand as an addend as many times as is indicated by the multiplier. The result of a multiplication operation is called the *product*. For example, 15 x 6 = 90. In essence, to multiply 15 by 6 means to add 15 to itself 6 times: 15 + 15 + 15 + 15 + 15 + 15 = 15 x 6 = 90. 15 is the multiplicand; 6 is the multiplier; and 90 is the product. Multiplication is sometimes referred to as the *short cut of addition*. In the given example, 15 and 6 are said to be the factors of the product 90. Instead of the multiplication sign x (read 'times'), a point - placed half way between the factors - is usually used. For instance, $15 \cdot 6 = 90$.

Division is an arithmetic operation, also initially defined in terms of repeated subtraction, by which the result of two quantities or numbers is calculated. Division is the inverse, or opposite, operation of multiplication, just as subtraction is the inverse operation of addition. In the case of division, the product and one of the factors are known. Therefore, what we are looking for is the other factor. Hence, *division is the finding of one of the factors by a given product and factor*. The given product is called the *dividend*; the given factor is called the *divisor*; and the unknown factor, which is the result, is called the *quotient*. For example, 54 ÷ 6 = 9; or $^{54}\!/_{6} = \dfrac{54}{6} = 9$. Here, 54 is the dividend; 6 is the divisor; and 9 is the quotient. The sign (÷) is usually used to indicate a division operation. Sometimes either a slash (/) or horizontal bar (–) is used instead, as in the preceding example.

Besides these four basic operations, there are also two other important operations that you will have to frequently deal with in the course of your studying and using mathematics. One is *raising a number or quantity to a power*. To raise a number to a whole (second, third, fourth, etc.) *power* means to repeat it as a factor two, three, four, etc. times. The second power is called a *square*, and the third power is a *cube*. For example, $2^3 = 2 \cdot 2 \cdot 2 = 8$. 2 is the *base* of the power; 3 is the power. Sometimes the word *exponent* is used instead of the word *power*. 8 is the answer or result of the operation. The expression 2^3 may be read as 2 *raised to the third power (or two raised to the power of three)*.

The other operation is *extraction of a root*. To *extract a root* means to determine the value of the root of a number or quantity. It is finding a number that when raised to the power of the given or indicated root gives or equals to the number or quantity under the root. The root of a number raised to the second power is otherwise known as its *square root*; the root of a number raised to the third power is otherwise known as its *cube root*; and the root of a number raised to the fourth power is its *fourth root*; and so on.. For example, in the expression $8^2 = 8 \cdot 8 = 64$, *8* is the *square root* of *64*. That is, $\sqrt[2]{64} = 8$ (reads "*the square root of 64 is equal to 8*"). In the expression $2^3 = 2 \cdot 2 \cdot 2 = 8$, 2 is the cube root of 8. That is, $\sqrt[3]{8} = 2$ (reads "*the cube root of 8 is equal to 2*").

In the example given above, 64 and 8 are the numbers under the roots; and 2 and 3 are the *indices* or *powers* (sometimes called *exponents*) of the root. It should be noted that in the symbol of a square root, as a rule, the power of the root is omitted. For example, it is generally accepted to write the expression for the square root of 9 ($\sqrt[2]{9}$) only as $\sqrt{9}$, without the 2. As you already know from previous classes, $\sqrt{9} = 3$.

The concept of extracting a root is often confusing to some pupils; hence this definition. A *root* is a number or quantity that, when multiplied by itself a certain number of times, yields a given number or quantity. For example, the fifth root of 32 is equal to 2. That is, 2 is the number or quantity that is multiplied by itself five times to yield the product 32. We write it like this: $\sqrt[5]{32} = 2$. This means $2^5 = 2 \cdot 2 \cdot 2 \cdot 2 \cdot 2 = 32$. Let us repeat that 2 is the root which, when multiplied by itself five times (in other words, when raised to the fifth power), yields the number under the root (or 32, in our case); and 5 is the power of the root. In general, *the expression or formula for the nth root of m is equal to p* is written like this: $\sqrt[n]{m} = p \rightarrow p^n = m$.

Other basic definitions are given below:

Prime numbers (or *prime integers*) are natural numbers, except 1, which are divisible only by themselves and 1. Examples of prime numbers less than 30 are: 2, 3, 5, 7, 11, 13, 17, 19, 23, and 29. The remaining other numbers which are not prime numbers are called *composite numbers*. They are integers that can be factorized into two or more other integers. Any composite number can be uniquely represented in the form of a product of *prime factors*. For example:

$50 = 2 \cdot 5 \cdot 5 = 2 \cdot 5^2$;
$75 = 3 \cdot 5 \cdot 5 = 3 \cdot 5^2$;
$100 = 2 \cdot 2 \cdot 5 \cdot 5 = 2^2 \cdot 5^2$;
$500 = 2 \cdot 2 \cdot 5 \cdot 5 \cdot 5 = 2^2 \cdot 5^3$.

The *greatest common divisor (GCD)* of several numbers is the greatest natural number by which all of these numbers can be divided.

The least common multiple (LCM) of several numbers is the smallest natural number which is divisible by all of these numbers.

There are two basic rules in finding the GCD and LCM:

1. *In order to find the GCD of two or more numbers, we determine the prime factors of each number; and then we calculate the product of the common (only the common) prime factors, taking each of them with the lowest exponent present.*

2. *In order to find the LCM of two or more numbers, we determine the prime factors of each number; and then we calculate the product of all (we emphasize all) of the prime factors, taking each of them with the highest exponent present.*

For example, let us find the GCD and LCM of the numbers 72 and 90:

a) We find the *prime factors* of the number 72. $72 =$
$$= 2 \cdot 2 \cdot 2 \cdot 3 \cdot 3 = 2^3 \cdot 3^2;$$

b) We find the *prime factors* of the number 90. $90 =$
$$= 2 \cdot 3 \cdot 3 \cdot 5 = 2 \cdot 3^2 \cdot 5;$$

c) We find the GCD of the numbers 72 and 90 $=$
$$= GCD\ (72, 90) = 2 \cdot 3^2 = 18;$$

d) We find the LCM of the numbers 72 and 90 $=$
$$= LCM\ (72, 90) = 2^3 \cdot 3^2 \cdot 5 = 360.$$

Note: In (c), the prime factor *5* is not used because it is not *common* to both (a) and (b); and we used the lowest exponent of each *common* prime factor. In (d), we used the highest exponent of each prime factor in both (a) and (b).

1.2   The Order of Arithmetic Operations

In performing arithmetic operations, the following order is observed as a rule.

1. The operations enclosed in brackets should be performed first. After that the operations should be carried out in the order of multiplication and division, and then followed by addition and subtraction. It is usually helpful to remember *"My Dear Aunt Sally"* or *"MDAS"*, as a clue for multiplication, division, addition, and subtraction. It may also be easier for others to remember *BODMAS*, which stands for *bracket of division, multiplication, addition, subtraction.* For example: $5 + (3 + 6) \cdot 4 - 2 = 5 + 9 \cdot 4 - 2 = 5 + 36 - 2 = 39$.

2. If the expression enclosed in brackets likewise contains brackets, then *the operations in the internal brackets are carried out first.* For example:

$$4 + 3 \cdot (15 - 20 \div (3 \cdot 2 - 4)) =$$
$$= 4 + 3 \cdot (15 - 20 \div (6 - 4))$$
$$= 4 + 3 \cdot (15 - 20 \div 2)$$
$$= 4 + 3 \cdot (15 - 20 \div 2)$$
$$= 4 + 3 \cdot (15 - 10)$$
$$= 4 + 3 \cdot 5 = 4 + 15 = 19.$$

It is important to remember that the order of performance of arithmetic operations is carried out from the *second stage* to the *first stage*, where *second-stage* operations are *multiplication and division*; and *first-stage* operations are *addition and subtraction*.

1.3   Signs of Divisibility

Sign of divisibility by 2. A number that can be divided by 2 (that is, without a remainder) is called an *even number*. Even numbers include the digits 0, 2, 4, 6, 8, and any number ending with them. In other words, *a number is divisible by 2 if its last digit is even or zero*. A number that cannot be divided by 2 is called an *odd number*. The first five odd numbers are 1, 3, 5, 7, and 9. A number that ends with any of these digits is not divisible by 2. For example, the number 15846 is divisible by 2, since the last digit is 6, an even number. The number 48351 is not divisible by 2, since the last digit is 1, an odd number. Explain why the number 5310 is divisible by 2.

Sign of divisibility by 4. *A number is divisible by 4 if its last two digits are zeros, or if its last two digits form a number that can be divided by 4*. Thus the number 98700 is divisible by 4; and 5835 is not divisible by 4, since 35 cannot be divided by 4. The number 916 is divisible by 4 because 16 can be divided by 4.

Signs of divisibility by 3 and by 9. *By 3 can be divided only those numbers for which the sum of their digits are divisible by 3. Likewise by 9 can be divided only those numbers for which the sum of their digits are divisible by 9*. The number 54321 is divisible by 3 and not divisible by 9, since the sum of its digits is equal to 5 + 4 + 3 + 2 + 1 = 15, which is divisible by 3. The number 7458093 can be divided by both 3 and 9, since the sum of its digits (36) is divisible by both 3 and 9.

Sign of divisibility by 5. *Those numbers can be divided by 5 the last digit of which is either zero or 5*. The number 930 is divisible by 5, since the last digit is 0. The number 246 is not divisible by 5 because the last digit is 6.

Sign of divisibility by 6. *A number can be divided by 6 if it is divisible simultaneously by 2 and by 3*. The number 584 is not divisible by 6. Though it is divisible by 2, it cannot be divided by 3. The sum of its digits (5 + 8 + 4 = 17) cannot be divided by 3. However, the number 348 is divisible by 6, since it can be divided by both 2 and 3.

Signs of divisibility by 10, 100, and 1000. *By 10 can be divided only those numbers the last digit(s) of which is (are) zero(s); by 100 can be divided only those numbers the last two (or more) digits of which are zeros; and by 1000 can be divided only those numbers the last three (or more) digits of which are zeros*. The number 2500 is divisible by 10 and 100, but not by 1000. The number 135000 is divisible by 10, by 100, and by 1000.

1.4   The Natural Number Series

When we talk about the natural number series, we are referring to the natural sequence of numbers. In mathematics, the need frequently arises to generalize or repeat partially known concepts or facts. The study of mathematics in the sixth grade is no exception. This is necessary to strengthen your understanding of already known principles and to help you conceptualize the laws and rules to which operations with both natural and fractional numbers are subjected.

Number is one of the basic and more frequently used concepts in mathematics. Numbers that are used in counting objects are known as *natural numbers. A natural number is any of the positive integers, such as 1, 2, 3, 4, and so on, which can be defined in terms of its quantity and position on the natural number line.* In previous classes you have had relatively adequate knowledge and experience with numbers and operations with them. You are able to arrange numbers in the order of their values to get the following sequence: 1, 2, 3, 4, 5, 6, 7, 8, 9, 10, 11, ...(and so on). It is possible to continue this sequence, writing the succeeding numbers. Such a sequence of numbers is known as *the natural number series.* It is necessary to know the following concepts about the natural number series.

The natural number series begins from 1. 1 is the least number on the natural number line. The largest natural number does not exist. After every number on the natural number line follows a strictly defined natural number.
Every number on the natural number line, except 1, is likewise preceded by a strictly defined natural number.

Natural numbers are used not only for counting, but also to show in which sequence it is expedient to consider objects. For examples, if we were to count some objects in order, and the last of them would be five, for instance, then it means that all of the objects are five.

However, there are also cases in which it is necessary to refer to objects according to their position in a sequence. This brings into focus the concept of cardinal and ordinal numbers. A *cardinal number* is *a number denoting quantity but not order in a group.* An *ordinal number* is one denoting relative position in a sequence, such as *first, second, third, fourth, fifth,* and so on.

1.5   The Concept of Place Value in the Writing and Reading of Natural Numbers

The following ten different digits are used for the writing of natural numbers in our base ten numeration system: 0, 1, 2, 3, 4, 5, 6, 7, 8, and 9. The value of the digits in the writing of a multi-digit number depends on the position of each digit in the number. This carries us back to the concept of *place value* that you studied in the 5th or 4th grade. If we move or shift a digit in a given number by one place to the left, then its value is increased by ten times; and if we shift a digit in a number by one place to the right, then its value is decreased by ten times. For example, the number 44337 has 40 thousands, 4 thousands, 3 hundreds, 3 tens, and 7 units. In this number, the value of 3 hundreds is ten times greater than the value of 3 tens. Analogically, the value of 4 thousands is ten times less than the value of 40 thousands. This represents an example of the notion of place value and of shifting a digit in a given number by one place either to the left or to the right. *Explain what happens when we shift a digit in a given number by two or more places either to the left or to the right. Give some examples.*

In the generally accepted way of writing and reading numbers in our numeration system, the number 10 has an important significance. Since we use the base-ten system, it is well known that *ten units make up one group of ten; ten groups of ten make up one group of hundred; ten groups of hundred make up one group of thousand;* and so on. In essence, every ten units of a lower rank make up one unit of a higher rank. That is why it is said that we use *decimal system of counting.* Since in the writing of numbers the place or position of a digit is of great importance, then it can still be said that we are using *positional decimal system of counting*. In this system, 1, 10, 100, 1000, and so on, are called *ranked units*. To symbolize the absence of that or another rank, a special symbol - the digit *zero* - is used. Zero is not a natural number. It indicates either an absence of a natural number, or a point of reference where counting begins.

For the convenience of reading a large number, we break it up from right to left into classes with three digits in each. The first three digits from the right make up the first class; the following three digits – the second class; and so on. See Table 1.1 below. The last class in a multi-digit number may have three, two, or one digit. For instance, the number 53,606,445 has three classes: 53 millions, 606 thousands, and 445 units. It is possible to write it in the form of a sum of ranked addends: 53, 606, 445 = 53,000,000 + 606,000 + 445 = (50,000,000 + 3,000,000) + (600,000 + 6,000) + (400 + 40 + 5). The digit 5 is found twice in this number, likewise the digits 6 and 4. The place value (or the value of the place) of the first 5 is *five units*, and that of the second 5 – *fifty millions*. Similarly, the place value of the first 6 is *six thousands*, and that of the second 6 – *six hundred thousands*. The place value of the first 4 is *forty (or four tens)*, and that of the second 4 – *four hundreds*. Here the first 5 is generally accepted to mean the first 5 from the right of the number. This also applies to the other digits (6 and 4). The digit 3 is found only once in the number; and its place value is *three millions*. The digit 0 signifies that the rank of *tens of thousands* is absent in the number.

In the 4th grade you might have studied natural numbers up to a billion and can read any nine-digit number. It is well known that the largest nine-digit number is 999, 999, 999. The least ten-digit number that follows it is *one billion,* (or 1,000,000,000*)*. As we already know that the largest natural number does not exist, there are numbers that are greater than a billion. For instance, the population of the whole world is estimated to be 7 billion. Also the distance (measured in meters) from the earth to the sun and many other planets is expressed in numbers greater than a billion.

Classes	Fifth class			Fourth class			Third class			Second class			First class		
	Trillions			Billions			Millions			Thousands			Units		
Name of Ranks	Hundreds of Trillion	Tens of Trillion	Trillions	Hundreds of Billion	Tens of Billion	Billions	Hundreds of Million	Tens of Million	Millions	Hundreds of Thousand	Tens of Thousand	Thousands	Hundreds	Tens	Units

Table 1.1. Classes and ranks of numbers

The class of *billions* also has three ranks just as the other classes. One unit of the rank of billion is 1,000,000,000; two units of billion is written 2,000,000,000; three units of billion is 3,000,000,000; and so on. Ten of such units of billion make up a new ranked unit written as 10,000,000,000 (or ten billions). Another ten of a group of ten billions make up a new ranked unit written as 100,000,000,000 (hundred billions). The class of *trillions* follows after the class of billions. It should be noted that ten groups of a hundred billions is equal to *one trillion* (that is, one unit of the rank of trillion), written as 1, 000,000,000,000.

The next classes of numbers higher in value than trillions are *quadrillions* and *quintillions*. Trillions, quadrillions, and quintillions are very rarely used. They are usually written in the form of scientific notation with which you will get acquainted in higher classes. Meanwhile, you can concentrate only on the first four classes in the Table 1.1 above and pay attention to the names of their respective ranks. *How well do you read each of the following numbers: 952,401,893,643; 708,453,126; and 45,123,791?*

1.6 Roman Numerals and Number Bases

Sometimes another method is used in the writing of numbers. Such symbols of numbers are usually seen on the dials of or faces of some watches, measuring instruments, and in the numeration of chapters of books. For the writing of these of these numbers they make use of the digits of the ancient *Roman system of counting or numeration*. The *Roman numeration system* is an example of *non-positional system of counting*. Only seven symbols, having the following values, are used in it:

I	V	X	L	C	D	M
1	5	10	50	100	500	1000

The remaining numbers are written, repeating the same symbols by defined rules. Some of the basic rules governing the use of Roman numerals are as follows:

1. Any one symbol must not be repeated more than three times in succession. In order to read a number, expressed by repetition of the same symbols, their values are added up. For example,

II = 2;
XXX = 30.

2. Addition of symbols is used when the symbol less in value stands after the greater symbol. For example:

XI = 10 + 1 = 11;
XVI = 10 + 5 + 1 = 16.

3. If the symbol less in value stands before the greater, then the value of the smaller symbol is subtracted from the value of the greater. For example:

IV = 5 − 1 = 4;

$IX = 10 - 1 = 9;$
$XL = 50 - 10 = 40;$
$CMXC = (1000 - 100) + (100 - 10) = 900 + 90 = 990;$
$MLMXLVI = 1000 + (1000 - 50) + (50 - 10) + 5 + 1$
$$= 1000 + 950 + 40 + 6 = 1996.$$

As you can see, the writing and reading of numbers in this system of counting is more complex than in the decimal system. Therefore, it is not used so often.

1.6.1   Operations with Number Bases

Base Two System of Writing Numbers

There exist other systems of writing numbers. For example, there is the *binary (or base two) system, in which only two symbols or digits (0 and 1) are used.* This system finds wide application in computer technology. In essence, the *binary system or notation is a number system having a base of two in which numbers are expressed in sequences of the digits 0 and 1, represented electronically as "off" and "on".* Some binary numbers are 11001101, 101010011, and 111.0001.

Some examples of converting from binary notation to the base-ten system are as follows:

(a) $110101001_2 = 1 \times 2^8 + 1 \times 2^7 + 0 \times 2^6 + 1 \times 2^5 + 0 \times 2^4 + 1 \times 2^3 + 0 \times 2^2 + 0 \times 2^1 + 1 \times 2^0$
$$= 1 \times 2^8 + 1 \times 2^7 + 1 \times 2^5 + 1 \times 2^3 + 1 \times 2^0$$
$$= 256 + 128 + 32 + 8 + 1 = 425;$$

(b) $1101.101_2 = (1 \times 2^3 + 1 \times 2^2 + 0 \times 2^1 + 1 \times 2^0) +$
$$+ (1 \times 2^{-1} + 0 \times 2^{-2} + 1 \times 2^{-3})$$
$$= (1 \times 2^3 + 1 \times 2^2 + 0 \times 2^1 + 1 \times 2^0) +$$
$$+ ((1 \times \frac{1}{2}) + (1 \times \frac{1}{8}))$$
$$= (8 + 4 + 0 + 1) + (\frac{1}{2} + \frac{1}{8})$$

$$= (13) + (\tfrac{5}{8})$$
$$= 13\tfrac{5}{8}.$$

<u>Note</u>: $2^{-1} = 1/2^1 = \tfrac{1}{2}$; $2^{-3} = 1/2^3 = \tfrac{1}{8}$; any number raised to the power of zero is equal to 1; for instance, 2 raised to the power of 0 is equal to 1 → $2^0 = 1$. Analogically, $3^0 = 1$; $4^0 = 1$; $x^0 = 1$; $y^0 = 1$; and so on.

Besides the binary system, we have the following systems of number bases: *base-three system, in which only three digits are used: 0, 1, and 2; base-four system, in which only four digits are used: 0, 1, 2, and 3; and so on.* Examples of conversion of numbers from these number bases to our base-ten system are given below.

Base Three System of Writing Numbers

(a) $2101_3 = 2 \times 3^3 + 1 \times 3^2 + 1 \times 3^0$
$\qquad = 2 \times 27 + 1 \times 9 + 1 \times 1$
$\qquad = 54 + 9 + 1$
$\qquad = 64;$

(b) $1012.012_3 = (1 \times 3^3 + 1 \times 3^1 + 2 \times 3^0) + (1 \times 3^{-2} + 2 \times 3^{-3})$
$\qquad\qquad = (1 \times 27 + 1 \times 3 + 2 \times 1) + ((1 \times \tfrac{1}{9}) + (2 \times \tfrac{1}{27}))$
$\qquad\qquad = (27 + 3 + 2) + (\tfrac{1}{9} + \tfrac{2}{27})$
$\qquad\qquad = (32) + (\tfrac{5}{27})$
$\qquad\qquad = 32\tfrac{5}{27};$

<u>Note:</u> $3^{-2} = 1/3^2 = \tfrac{1}{9}$; $3^{-3} = 1/3^3 = \tfrac{1}{27}$.

Base Four System of Writing Numbers

(a) $1302_4 = 1 \times 4^3 + 3 \times 4^2 + 2 \times 4^0$
$\qquad = 1 \times 64 + 3 \times 16 + 2 \times 1$
$\qquad = 64 + 48 + 2$

$$= 114;$$

(b) $\qquad 0123.103_4 = (1 \times 4^2 + 2 \times 4^1 + 3 \times 4^0) + (1 \times 4^{-1} + 3 \times 4^{-3})$

$$= (1 \times 16 + 2 \times 4 + 3 \times 1) + (\frac{1}{4} + \frac{3}{64})$$

$$= (16 + 8 + 3) + (\frac{19}{64})$$

$$= (27) + (\frac{19}{64})$$

$$= 27\frac{19}{64};$$

Note: $4^{-1} = 1/4^1 = \frac{1}{4}$; $4^{-3} = 1/4^3 = \frac{1}{64}$.

Base Five System of Writing Numbers

(c) $3142_5 = 3 \times 5^3 + 1 \times 5^2 + 4 \times 5^1 + 2 \times 5^0$

$$= 3 \times 125 + 1 \times 25 + 4 \times 5 + 2 \times 1$$

$$= 375 + 25 + 20 + 2$$

$$= 422;$$

(f) $123.042_5 = (1 \times 5^2 + 2 \times 5^1 + 3 \times 5^0) + (4 \times 5^{-2} + 2 \times 5^{-3})$

$$= (1 \times 25 + 2 \times 5 + 3 \times 1) + ((4 \times \frac{1}{25}) + (2 \times \frac{1}{125}))$$

$$= (1 \times 25 + 2 \times 5 + 3 \times 1) + (\frac{4}{25} + \frac{2}{125})$$

$$= (25 + 10 + 3) + (\frac{22}{125})$$

$$= (38) + (\frac{22}{125})$$

$$= 38\frac{22}{125}.$$

Note: $5^{-2} = 1/5^2 = \frac{1}{25}$; $5^{-3} = 1/5^3 = \frac{1}{125}$.

1.7 Rounding Off of Numbers

Let us imagine that as a result of a population census, it was determined that 985,756 people live in Monrovia; but already in several months or so this number may change. After some time, the digits in its *units* and *thousands* position would certainly not remain the same. Consequently, it makes sense to provide an estimate of how many people live in Monrovia. This requires census numerators to *round off* or *approximate* the value of the said population *to the nearest million*, for example. Numbers are rounded off using the sign " ≈ "; therefore, we can write the number 985,756 rounded off to the nearest million as 985,756 ≈ 1,000,000.

Rounding off of a number is the replacement or substitution of that number by its approximate value. The new number that is formed is said to be *rounded off.* In rounding off a number to a *definite rank*, all of its digits from the right of that *rank* are dropped and replaced by zeros. If, by rounding off a number, the digits in its *units* and *tens* positions were dropped and replaced by zeros, then it is generally accepted to say that the rounding off was done *to the nearest hundreds*. If they rounded off a number *to the nearest thousands*, then the last three digits from the right of that number were dropped and replaced by zeros. If, likewise, by rounding off a number the digits in its *units, tens, hundreds, and thousands* positions were dropped and replaced by zeros, then it is said that the rounding off was performed *to the nearest tens of thousands*, and so on. Thus it is easy to see that *the digit of the rank to which a number is rounded off* is important and should be given as much attention as possible in order to avoid a mistake. *This digit is either maintained as is or increased by 1, depending on the closest digit before it from the right.* If the closest digit from the right is less than 5, then it is maintained as is; but if the closest digit from the right is 5 or greater than 5, then the digit of the rank to which the number is required to be rounded off is increased by 1, while all other digits from its right are dropped and replaced by zeros.

Therefore, two basic rules apply in the rounding off of numbers:
The first step is to determine the position of the digit of the rank to which a number is required to be rounded off. If the closest digit from the right of the *rank digit* is less than 5, then that *rank digit* is not changed (or maintained as is); and all other digits to its right are dropped and replaced by zeros. *In this case, after the rounding off operation, we have a resulting number that is less than the given number required to be rounded off. Such a rounding off operation is called a rounding off with deficiency.* Some examples are given below.

(a) Round off the number 165,342 to the nearest thousands: $165,342 \approx 165,000$;

(b) Round off the number 214,295,436 to the nearest tens of million: $214,295,436 \approx 210,000,000$;

(c) Round off the number 321,659,480,531 to the nearest hundreds of billion: $321,659,480,531 \approx 300,000,000,000$.

In each case, we can calculate how much the resulting rounded off number is less than the given number.

If the closest digit from the right of the *rank digit* is 5 or greater than five, then the *rank digit* is increased by 1; and all other digits to its right are dropped and replaced by zeros. *In this case, after the rounding off operation, we have a resulting number that is more than the given number required to be rounded off. Such a rounding off operation is called a rounding off with an excess or surplus.* Some examples are given below.

Round off the number 563,489 to the nearest hundreds: $563,489 \approx 563,500$;

Round off the number 835,096 to the nearest tens of thousand: 835,096 ≈ 840,000;

Round off the number 103,579 to the nearest tens: 103,579 ≈ 103,580.

In each case, we can calculate how much the rounded off number is more than the given number.

Let us consider another example. The number 1,723,856 rounded off to the nearest hundreds will be 1,723,900. If we round it off to the nearest tens, we will have 1,723,860; and if to the nearest millions, we will have 2,000,000.

Rounded off numbers are convenient to use for estimation of the results of operations with large numbers. For instance, regarding the product of the numbers 596 and 6102, it is possible to approximately equate it to the product of 600 and 6000, which is close to 3,600,000. In reality, the product of 596 and 6102 is equal to 3,636,792. In such a way, it is convenient to roughly check the results of numerical calculations.

1.8 An Introduction to Geometric Concepts

From the fifth grade or lower mathematics classes, you may be already acquainted with such concepts as a *plane, curve, straight line,* a *ray, segment, broken line,* and a *point.* All of these are *geometric figures.* The simplest or basic among them is the *point.* The others are formed with the help of a point.

For example, the points used in the numeration of this topic (1.8) or at the end of sentences give an idea of *a point.* Points are most often symbolized by capital letters of the Latin alphabet.

A tightly stretched thread or a solar ray passing through a narrow crack may serve as an example of a part of *a straight line.* An example of *a plane* may be the surface of a table, a window glass, a sheet of paper in your copy book, or the page on which you are presently reading. Geometric figures can be represented on a sheet of paper. By a sharp pencil, one is able to leave a point or draw a straight line on a sheet of paper. You already might have drawn other geometric figures including *squares, triangles, rectangles,* etc.

If an arbitrary point is taken on a straight line, then that point divides the straight line into two half straight lines, or rays. Consequently, *a ray is a half straight line, having a beginning.* If two arbitrary points are taken on a straight line, then the part of the straight line confined between them is called a *line segment*; and the two points are said to be its *ends.*

It is important to learn how to correctly construct or draw line segments and to measure their lengths. In order to draw a segment of a straight line, it is necessary to take a ruler and press it to a sheet of paper. Hold your pencil in the right or left hand (depending on which one you use) and gently press it to the edge of the ruler and the sheet of the paper, having bent it a little to the direction of movement and from yourself. A line segment, as a rule, is symbolized or designated by two capital letters, indicating its ends. See Fig. 1.1 (a) and Fig. 1.1 (b).

Fig.1.1. (a) Line segment CD.

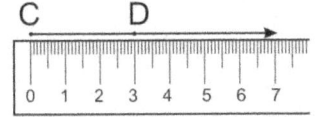

Fig.1.1. (b) Measuring line segment CD.

Every line segment has a definite length. In order to measure the length of the line segment represented in Fig. 1.1 (a), the zero on the ruler must coincide with the beginning of the line segment at point C; then the mark on the ruler, coinciding with the end of the line segment at point D, shows the length of the *line segment CD. Let each pupil measure a line segment CD of arbitrary length and tell the class the result of his or her measurement in either centimeters or inches.*

The length of a line segment can also be measured with the help of a pair of compasses. In order to do this, the legs of the pair of compasses should be accurately set at *points C and D (the beginning and end of the line segment);* after that, not changing the spread of the pair of compasses, one leg should be placed on *zero* on the ruler; then the second leg will show the length of line segment CD. *With the help of a pair of compasses, let each pupil measure a line segment CD of arbitrary length and tell the class the result of his or her measurement in either centimeters or inches.*

For measuring and transferring the value of the length of a line segment, it is possible to use a pair of *drawing compasses* as well as a pair of *measuring compasses.* In the former, one leg is in the form of a needle, and the other leg has a pencil at its end. In the latter, both legs have needles at their ends. It has already been explained above how, with the help of a ruler, the line segment CD was drawn and measured. Fig. 1.2 represents a transfer of the value of the length of the line segment CD with the help of a pair of compasses to another place. In this way, we have line segments MN and MP. It is quite obvious that the lengths of line segments CD, MN, and MP should be of the same value (equal to one another), since the length of segment CD has been transferred to MN and MP.

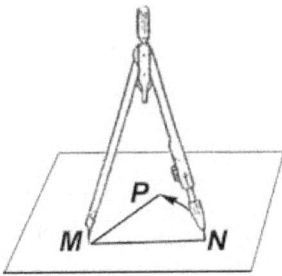

Fig. 1.2. Measuring and transferring the length of a line segment with the help of a pair of compasses.

Let us consider another line *segment PR,* with *point Q positioned between P and R.* Point Q divides PR into two line segments PQ and QR, as shown in Fig. 1.3. In other words, segment PR is made up of two segments PQ and QR, having a common point Q. Therefore, PR is a sum of PQ and QR; that is, PR = PQ + QR.

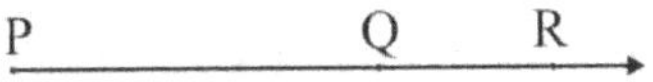

P   Q   R

Fig.1.3. Line segment PR, with a common point Q dividing it into two.

It follows from here that there is a way of *adding line segments.* Let us assume it is required to find the length of line segment ST as a sum of its component line segments. See Fig. 1.4.

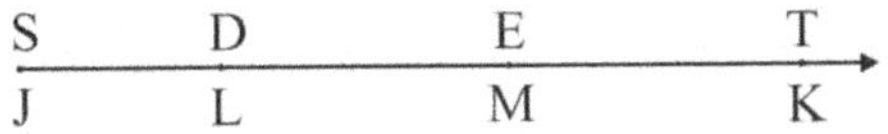

S   D   E   T
J   L   M   K

Fig. 1.4. Addition of line segments.

We first consider the ray ST. From its beginning at point S, with a pair of compasses, we measure and set aside the length of line segment JL. Likewise from point L we measure and set aside the length of line segment DE. Finally, from point E, we do the same

for line segment MK. Consequently, line segment ST is equal to the sum of the given line segments:
ST = SD + DE + EK = JL + DE + MK = JK.

Knowing the lengths of line segments, it is easy to establish a correlation between them. It does not matter what unit of measurement we use for the length of a line segment. For example, if JL = 3 inches, DE = 6 inches, LM = 6 inches, then JL < DE or DE > JL, DE = LM. If the lengths of line segments are unknown, then we can compare them by way of superposition of one segment upon the other with the help of a pair of compasses. Thus, we can see that JL is shorter than DE; and that DE is equal to LM, since their ends coincide.

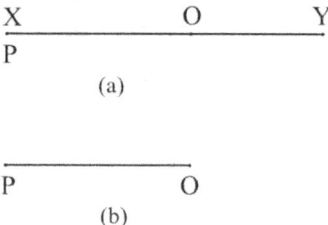

(a)

(b)

Fig. 1.5. The difference of two line segments.

If XY > PO (Fig. 1.5 (a)), then OY is equal to their difference, that is, OY = XY − PO. You should try to explain why this is true.

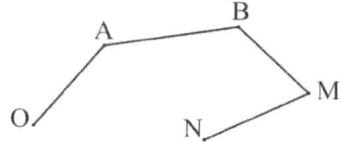

Fig. 1.6. A broken line segment ON.

In previous figures, line segments were placed on one straight line. But in Fig. 1.6, they are placed otherwise. They form so-called broken lines. A *broken line segment is a succession of line segments placed (not in a straight line) so that the end of the first segment is the beginning of the second, the end of the second is the beginning of the third; and so on.* The length of the broken line segment ON (Fig. 1.6) is equal to the sum of the lengths of line segments of which it is comprised. For example, the length of ON is equal to the sum of the broken line segments OA, AB, BM, and MN; that is, ON = OA + AB + BM + MN.

1.9   The Number Line. Comparison of Natural Numbers

Let us consider a straight line CD in Fig. 1.7 and mark on it an arbitrary point O. Point O is a point of reference and divides the straight line into two rays. From one side each ray is limited or confined to point O, which is its beginning. The rays so formed are OC (directed from right to left) and OD (directed from left to right). Rays OC and OD are said to be complementary. *Complementary rays are two rays lying on a straight line, having a common beginning and opposite directions.*

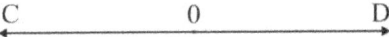

Fig. 1.7. A straight line depicting the concept of the number line.

Again, let us take another look at ray 0x in Fig. 1.8 (b). We take the segment OA in Fig. 1.8 (a) to be our *unit length*. That is, the length of segment OA is considered as equal to 1; and we make it to lie along the ray 0x from its beginning at point 0. In other words, we divide the entire ray 0x into unit lengths, each of which is equal to OA. We now take another point represented by B along 0x. We make sure that 0B = OA = 1. Further from point B, we again make the unit length OA to lie along the ray 0x. We now have another point, symbolizing the number 2. It is obvious to us that at 2, it takes two of the unit segment OA. If we continue this process, the given ray will acquire the appearance as represented in Fig. 1.8 (b). A ray in the form of a straight line having a beginning and defined direction together with a unit length is called a *number line*.

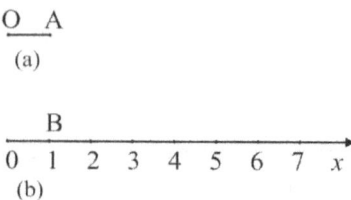

Fig. 1.8. A number line.

On a number line each point corresponds to a natural number. If point B corresponds to the number 1 and vice versa, then it is written like this: B (1). It is convenient to compare numbers with the help of a number line. *On the ray directed from left to right, each point lying to the right corresponds to a greater number; and each point lying to the left corresponds to a smaller number. In other words, given two points on a number line, the one lying on the right is greater than the one on the left.*

For the comparison of multi-digit numbers, it is expedient to be guided by the following rules:

If two natural numbers have different quantity of digits, the one having more digits is greater. For example, the number 102,034 > 98, 796.

If two natural numbers have the same quantity of digits, the one with a greater digit in the highest rank is greater. For example, the number 957 > 875.

3. If the *digits in the highest ranks of* any two natural numbers (having the *same quantity of digits) are equal, then the digits in their next higher ranks are compared. For example:*

$$(a) \; 6543 < 6643;$$
$$(b) \; 5512 > 5502;$$
$$(c) \; 4971 < 4973.$$

For the measuring of the numerical values of different physical quantities in many devices or instruments, they use one strip or another with graduated units or points which correspond to definite values of the quantity being measured. Such a strip is called *a scale*. A scale may be either *rectilinear (*also called *straightforward) or curvilinear*. An ordinary ruler and thermometer are examples of *rectilinear scales*. The dial of a clock, the scales used in measuring weights, and an automobile speedometer are examples of *curvilinear scales*. Every graduated unit or point on a scale corresponds to a definite quantity. For instance, the basic graduated unit on a ruler corresponds to 1mm ($\frac{1}{10}$ of a centimeter), 0r $\frac{1}{2}$ of an inch; on a thermometer, each graduation is 1 degree; on a watch or clock, each graduation is 1 minute or 1 second; and so on.

1.10   Rational Numbers

In elementary mathematics, emphasis is placed on the following sets of numbers: natural numbers; whole numbers; rational numbers; irrational numbers; and real numbers (consisting of both rational and irrational). Standard notations or symbols of these number sets are given below:

N = {1, 2, 3, 4, 5, 6, 7, ...} – is a number set consisting of all ***natural numbers***; that is, the set of all ***positive whole numbers***.

Z = {0, ±1, ±2, ±3, ±4, ±5, ±6, ±7, ...} – is a number set consisting of all *whole numbers*; that is, the set of all natural numbers and all numbers opposite to them, including zero.

Q = $\left\{\frac{m}{n}\right\}$, where $m \in Z$, $n \in Z$, and $n \neq 0$; - is a number set consisting of all *rational numbers*; that is, the set of all *positive numbers* (including *whole numbers* and *fractional numbers*), all negative numbers (including *whole numbers* and *fractional numbers)*, and *zero*.

R = {x}, where $-\infty < x < +\infty$ - is a number set consisting of all *real numbers x* (*rational and irrational numbers* combined);

A *rational number* $\frac{m}{n}$ may be formally defined as either a *whole number,* or it is possible to be represented in the form of a *finite* or *periodically infinite decimal fraction* (see also the section 4.9 on decimal approximations of common fractions in this book). Examples of rational numbers are −7, −3.5, −1.272727..., 0, 5, 0.85, 6, 0.333..., etc.

An *irrational number* is represented by a non-periodic infinite decimal fraction. The numbers $\sqrt{2}$, 5.30146895..., 1.5454780125867..., etc. are examples of irrational numbers.

In previous sections of this chapter we discussed and learned only about natural numbers which are positive numbers. The study of mathematics also involves our dealing and operating with both positive and negative numbers as introduced. Concepts about the different number sets, including rational numbers in particular, and efficient operation with them form part and parcel of the structure of prerequisite knowledge and understanding required to continue the successful study of mathematics as you advance on a higher rung of your academic and educational pursuits.

As noted earlier, *negative* and *positive numbers* are the main component elements in the set of rational numbers.

The writing or reading of numbers with a ***plus*** or ***minus sign*** is very important in telling us about the direction on a scale such as the thermometer and very much so on the coordinate number line. For instance, it is not sufficient to tell or inform a person that the temperature is $7\,^0C$ if that person is interested in knowing whether or not there is frost outdoors on a winter day or night. Normally, for someone to understand whether or not there is frost, it is necessary to either say $7\,^0C$ *below zero* or $7\,^0C$ *above zero*, since $7\,^0C$ *below zero* (or $-7\,^0$C) and 7 ^{0}C *above zero* (or $+7\,^0$C) are two unique readings that are in opposite directions on the thermometer scale.

Moreover, "below zero" or "above zero" is usually added because temperature may change from zero to opposite directions, just in the same manner numbers on the coordinate straight line lie in opposite directions from the zero.

A temperature may increase (that is, the mercury in a thermometer may move from zero upward), or decrease (that is, the mercury in a thermometer may move from zero downward). Consequently, it is accepted to write temperatures as $-6\,^0$C, $+2\,^0$C, $-1\,^0$C, or $+3\,^0$C. As a rule, however, if a temperature shows a positive reading (above zero), then the *plus* sign is usually omitted and instead of $+9\,^0$C for instance, it is accepted to simply write 9 ^{0}C; or instead of $+18\,^0$C, it is customary to simply write 18 ^{0}C.

In this way, for the writing of temperatures or numbers greater than zero, we use numbers which are called *positive numbers*; and for the writing of those less than zero, we use numbers which are called *negative numbers*. It should be noted that the number *zero* belongs to the set of *whole numbers*; and it is customary to write *zero* without a sign.

Negative numbers are used not only for the measurement of temperatures. For instance, when we talk about property and debt, profit and loss, about past and future times, payment to a cashier's office and payment from there, the height of an object above the level of sea and other reference point, we make use of positive and negative numbers.

1.11   Coordinate Straight Line

A straight is drawn in Fig.1.9. In the middle of the straight line from right to left we mark a point and mark it as 0 (*zero*). *Zero* is given as the reference point, or the point of origin. We mark the points M, N, P, R, S, and T on the straight, where:

(a) M lies by 10 unit segments to the left of 0;
(b)   N lies by 6 unit segments to the left of 0;
(c) P lies by 2 unit segments to the left of 0;
(d)   R lies by 2 unit segments to the right of 0;
(e) S R lies by 6 unit segments to the right of 0; and
(f) T R lies by 10 unit segments to the right of 0;

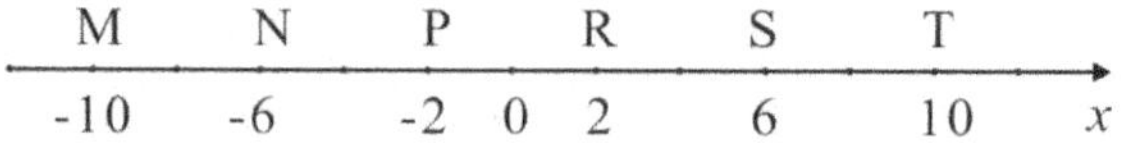

Fig. 1.9. A coordinate straight line.

Instead of the words "to the right of *zero*" and "to the left of *zero*" we use signs "+" and "−". As you already know, numbers located to the right of the zero are written with a *plus sign*, such as +1 (plus one), +2 (plus two), +3 (plus three), +4 (plus four), +5 (plus five), and so on; numbers located to the left of the zero are written with a *minus sign,* such as −1 (minus one), −2 (minus two), −3 (minus three), −4 (minus four), −5 (minus five), and so on. The direction to the right from point *zero* (0) is considered as the *positive direction*, while the direction to the left from point *zero* (0) is considered to be the *negative direction.* See Fig. 1.9. The positive direction on the straight line is shown with an arrow because counting increases in that direction. *A number showing the position of a point on a straight line is called the coordinate of that point.*

Point P in Fig. 1.9 has the coordinate −2 which may be written or symbolized as P(−2); point R has the coordinate 2, written as R(2); point N has coordinate −6, written as N(−6); point S has coordinate 6, written as S(6); point M has coordinate −10, written as M(−10); and point T has coordinate 10, written as T(10). *Any straight line on which there is a reference point (or origin point), a given unit segment (length of which is taken as equal to 1), and an indicated positive direction is called a coordinate straight line.*

The distance between two points x_1 and x_2 (as N and S, for example in Fig. 1.9 on a coordinate straight line is equal to the module of the difference of their coordinates:

$$|NS| = |x_1 - x_2| = |N - S| = |-6 - 6| = |-12| = -(-12) = 12.$$

Points P and R (Fig. 1.9) with coordinates -2 and 2 are equidistant from point 0 and lie on different sides from it in opposite directions. *Such numbers as -2 and 2, -6 and 6, -10 and 10 are called opposite numbers*. It is obvious that zero does not have an opposite number; it is therefore expedient to say that it is opposite to *itself.*

1.12   The Module of a Number

In mathematics, a *module* is defined as an absolute magnitude or absolute value. The *module of a number is the absolute value of that number*. Let us look at an example in order to better conceptualize that definition.

Two motor cyclists departed and traveled into opposite directions from a central bus station. Such description presents to us a similar picture of a *coordinate straight line* where the central bus station is our *point of origin* (*or point of reference*); and we can imagine the two motor cyclists as moving into opposite directions from the *zero* point. The first motor cyclist traveled 3 miles from the *zero* into the positive direction (i.e. to the right of the zero). The coordinate here on our described *coordinate straight line* will be +3. The second motor cyclist traveled 5 miles from the zero into the negative direction (i.e. to the left of the zero). The coordinate here on said *coordinate straight line* will be −5.

Problem 1.   Which of the two motor cyclists had traveled a greater distance?

Solution.   Comparing the distances traveled by the two motor cyclists, we are interested only in the value or magnitude of each traveled distance, irrespective of the direction into which it was traveled. A traveled distance is always measured as a *positive* number. Therefore, instead of −5, we will take the opposite number to it +5. The number +5 expresses the actual distance traversed by the second motor cyclist and, needless to say, is greater than +3. In such a case, we say that the number 5 is the **module** of −5. For the symbol of a module, two vertical lines are used.

So the module of the number −5 is equal to 5. It is written like this: $|-5| = 5$.

Problem 2. Identify or indicate the following points on a coordinate straight line: F(−7); G(−5.25); H(−3); I(−1.5); J(2.5); L(4); M(6); N(9). Find the *distance* of each point from the *point of origin*.

On a coordinate straight line, the distance from the *point of origin* to a point representing a number is called the *module* of that number. The following statements are true:

- *the module of a positive number is the same as that number;*
- *the module of the number zero is equal to 0;*
- *the module of a negative number is the number opposite to it;*

Consequently, the module *m* of a number may be represented by the expression:

$$|m| = \{m, \text{ if } m \geq 0;$$

$$|m| = \{-m, \text{ if } m < 0.$$

Examples:

(a) $|18| = 18;$

(b) $\quad |0| = 0;$

(c) $|-23| = -(-23) = 23;$

46

(d) $|-11.8| = -(-11.8) = 11.8;$

(e) $|0.99| = 0.99.$

1.13   Comparison of Rational Numbers

It is easy for us to compare natural numbers. For example: $9 > 7$, $3 < 10$, and $40 > 39$. Since the number zero is less than any positive number it appears confusing, if not difficult, for some of us to easily compare negative numbers. For example, Tombekai believes and strongly argues that $-17^0 > -8^0$ because 17 degrees of frost is colder than 8 degrees of frost. But Swen believes otherwise and argues that $-17^0 < -8^0$ because at the temperature of -8^0 it is warmer than at the temperature of -17^0. How can you resolve the arguments between the two boys? What can you say or show as a convincing proof?

Problem 1. Identify or represent the points W (−8) and V (−17) on a coordinate straight line. Which of them is greater?

Problem 2. Compare the coordinates of points M and N in Fig. 1.9. Which is smaller in value?

Problem 3. Compare the coordinate of point P and the number *zero* in Fig. 1.9. Which is greater?

Problem 4. Compare the coordinates of points M and R in Fig. 1.9. Which is greater?

Problem 5. Besides those already represented, mark six other points into both positive and negative directions from the zero in Fig. 1.9. Compare the coordinates of the points you have marked.

The following *rules* or statements are true when comparing rational numbers:

- *Of two numbers on a coordinate straight line, the one on the right is greater; conversely, of two numbers on a coordinate straight line, the one on the left is smaller.*

- *Every positive number is greater than zero; and every negative number is less than zero. It follows from here that every positive number is greater than every negative number.*

- *Of two negative numbers, that one is greater of which the module is smaller.*

Examples:

(a) $7.2 < 8.5$, since 8.5 is located to the right of 7.2 on the coordinate straight line;

(b) $6 > -11$, since -11 is located to the left of 6 on the coordinate straight line;

(c) $3 > 0$, since 3 is located to the right of 0 on the coordinate straight line;

(d) $0 > -50$, since -50 is located to the left of 0 on the coordinate straight line;

(e) $-7 > -19$, since the module of the number -7 is smaller than the module of the number -19;

(f) $-25 < -1$, since the module of the number -25 is greater than the module of the number -1.

1.14   Rectangular System of Coordinates

Problems which we are required to solve, making use of the rectangular system of coordinates, directly relate to finding a point on a plane. For example, let us suppose that the Brooks live in an apartment in a building located on X-Street in down town in the City of Monrovia. X-Street runs from west to east and is perpendicular to Y-Street, which runs from south to north. Situated to the left side of X-street towards the east, the building housing the Brooks' apartment is the fifth building from the intersection of X- and Y-Streets. We designate the intersection as point *zero*, which is our *point of origin* (or *point of reference*). Another family, the Jays, lives in an apartment located to the right side of Y-Street. Their apartment is in the seventh building *going towards the north* from the intersection of X- and Y-Streets. There is a supermarket around there which is *situated perpendicularly to the right* of the Jay's apartment building; i.e. *when facing north*. The supermarket is *the fifth building towards the east* from the apartment building of the Jays. Now, let us consider a problem.

Problem 1. How can we find the position of the supermarket?

Solution: Our task here is to find the position, or *coordinates,* of the supermarket. As easy as it is, it is obvious that the *position* of the supermarket can be found either by walking eastwards through the distance required to get to the *fifth* building from the Jays' apartment building; or by walking northwards through the distance required to get to the *seventh* building from the Brooks' apartment building. The description of the coordinates of the supermarket is shown in Fig. 1.10.

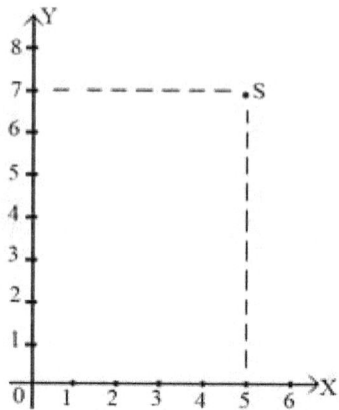

Fig. 1.10. The position of point S on a coordinate plane.

If we designate the supermarket by point **S**, then the position, or *coordinates*, of point S are given by the numbers 5;7 representing *x;y*. As we can see, the solutions to many practical problems lead to finding a point on a *coordinate plane*.

Problem 2.   Draw on a plane in your copybook two mutually perpendicular straight lines *0x* and *0y*, which intersect at point *0*. To determine the position of a point on a plane, divide it (the plane) into squares. On this plane draw a point P located from the straight line *0y* at the distance of 3 *unit segments* to the right, and from the straight line *0x*, at the distance of 4 *unit segments* upwards. See Fig. 1.11.

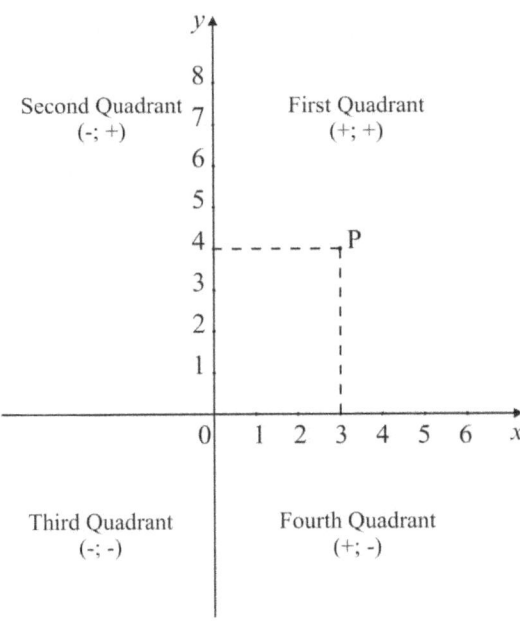

Fig. 1.11. Rectangular system of coordinates

The numbers 3 and 4, designating the position of point **P** on the plane, are called the *coordinates* of point P on the plane. The number 3 is called the *abscissa* of point P; and the number 4 is called the *ordinate* of point P (also see section 2.5.3, *Addition of Rational Numbers*). To designate the position of point P, it is customary to write P(3;4). *The abscissa of a point is always written as the first coordinate in the bracket; and the ordinate is the second coordinate. Mutually perpendicular straight lines with the help of which the position of a point (point P, for example) is determined on a plane are called coordinate straight lines or axes of coordinates.*

Mutually perpendicular straight lines 0x *and* 0y *with chosen direction on them, a common point of origin, and a unit segment are said to form what is called a* **rectangular system of coordinates**. A plane on which such system of coordinates are chosen or plotted is called a *coordinate plane.* The axes of coordinates divide the plane into four parts. Each part is called a *coordinate quadrant*. Numeration of the *first, second, third,* and *fourth* coordinate quadrants and the signs of coordinates according to these quadrants are shown in Fig. 1.11.

Application of *rectangular system of coordinates* on the plane is connected with the name of an outstanding French philosopher and mathematician Rene Descartes (1596-1650). Application of a system of coordinates allows the realization of connection between algebra and geometry and makes it easier for us to solve many problems of practical nature.

Chapter One: Questions to Test Your Understanding
A.
1. Define the following terms:

(a) Addend	(g) Multiplier
(b) Sum	(h) Product
(c) Difference	(i) Dividend
(d) Minuend	(j) Divisor
(e) Subtrahend	(k)Quotient
(f) Multiplicand	

2. In your own words explain why multiplication is considered as *the short cut of addition.*

3. What is meant by *raising a quantity to a power*? Give an example.

4. What is *extraction of a root*? Give an example.

5. In which order arithmetic operations are performed?

6. What signs of divisibility by the numbers 2, 4, 3 and 9 do you know?

7. What signs of divisibility by the numbers 5, 6, 100, and 1000 do you know?

8. What is a natural number?

9. Explain briefly the concept of place value as it relates to our base-ten numeration system?

10. Why it is said that we are using *positional decimal system of counting*?

B.
1. Write the following numbers in digits:
   (a) Five billions two hundred thousands nine hundreds and five;
   (b) Two millions three thousands and one;
   (c) Nine thousands and four.

2. Write the first eleven natural numbers.

3. A thousand is a number represented as one followed by four zeros.
   (a) What is a quadrillion?
   (b) What is a quintillion?

4. What are three basic rules that govern operations with Roman numerals? Give an example of each rule.

5. Write the following numbers in words:
   (a) 81,726,354;
   (b) 59,073;
   (c) 65,800.

6. Besides the decimal and Roman systems of counting, what other systems of writing numbers do you know?

7. What is the binary system? Give an example.

8. What is the base-four system? Give an example.

9. Considering your answers to questions 9 and 10, explain which principle forms the basis of writing natural numbers in the decimal system.

10. What is the largest seven-digit natural number?

11. What is the smallest nine-digit natural number that does not contain the digit zero?

12. What is the major difference between the writing of numbers in the decimal and Roman notations?

13. Briefly explain two basic rules used in rounding off numbers.

C.
1. Define the following terms:
   (a) Ray
   (b) Line segment
   (c) Complementary rays

2. Give some examples of a plane.

3. How is it possible to transfer the measurement of a line segment to another place?

4. Name two different types of *scales used in measuring instruments* and define them.

5. Give examples of the types of *scales that are used in measuring instruments*.

6. How can we compare two numbers with the help of a number line?

7. Give at least two important rules that are used to compare natural numbers.

8. Explain the following phrases:
   (a) Rounding off with a deficiency;
   (b) Rounding off with a surplus.

Chapter One: Problems and Exercises

1. Find the results of the indicated operations:
    (a) $345 + 23 + 9{,}506 + 54{,}321$;
    (b) $4{,}013 - 1{,}999$;
    (c) $78{,}095 \cdot 34$;
    (d) $16{,}284 \div 12$.

2. What is the value of 7 raised to the third power?

3. What is the value of 5^4?

4. What is the cube root of 125?

5. What is the value of the expression $\sqrt[4]{10{,}000}$?

6. What is x if the value of x raised to the second power is 64?

7. What is the fourth root of 81?

8. Determine the value of the given expression, observing the order of operations:
    $(5 + 11) \cdot 2 - 6 + 8 \div 4$.

9. Determine the value of the given expression, observing the order of operations:
    $4 \cdot (8 - 9 \div (3 \cdot 5 - 12)) + 7$.

10. Determine which of the given numbers can be divided by 2:
    (a) 9,851;
    (b) 1,502;
    (c) 963;
    (d) 1,284

11. Determine which of the given numbers can be divided by 4:
    (a) 8,502
    (b) 5,600

(c) 9,532
(d) 7,654

12. Determine which of the given numbers can be divided by 3:
    (a) 1,256
    (b) 3,617
    (c) 2,652
    (d) 1,090

13. Determine which of the given numbers can be divided by 9:
    (a) 1,458
    (b) 2,718
    (c) 5,436
    (d) 9,090
14. Determine which of the given numbers cannot be divided by 5:
    (a) 5,030
    (b) 6,785
    (c) 9,000
    (d) 9,051

15. Determine which of the given numbers cannot be divided by 6:
    (a) 5,436
    (b) 1,978
    (c) 1,239
    (d) 4,500

16. Determine which of the given numbers can be divided by 10, 100, and 1000:
    (a) 25,000
    (b) 905,000
    (c) 2,000
    (d) 5,000

17. There are 21 different flowers on the stall of a seller in a flower market. If the flowers are counted from left to right, a rose occupies the ninth place. Which will be the ordinal number of the rose if the flowers are counted from right to left?

18. What are the least four-digit and the greatest two-digit numbers? Find the difference between these numbers.

19. In a garage there were 14 Toyota buses. The number of Renault buses in the garage was 5 less than the number of Toyota buses. How many buses in all were there in the garage?

20. 120 pounds (lb) of mangoes, 35 lb of cassavas, 60 lb bananas, 150 lb of yams, 70 lb of butter pear, and 30 lb of sweet potatoes were bought and delivered to a school canteen. How many pounds of fruits were delivered more than vegetables? Cassava, yam, and sweet potato are considered as vegetables; mango, banana, and butter pear – as fruits.

21. Let us suppose that one of the natural numbers has been designated by the letter x. Write the three preceding numbers and the three succeeding numbers. Separated by a semi-colon, write the sequence in which x and the determined numbers appear. What would the natural sequence of x and the determined numbers be, for example, if x were 10?

22. Take any three consecutive natural numbers and add them. Observe (and be convinced) that the sum of such three numbers is divisible by 3. When this number is not divisible by 6?

23. Write the following numbers in words:
     (a) 25,368,147,908
     (b) 1,309,452,879
     (c) 813,456,207,912
     (d) 1,000,000,014

24. Write the following numbers in digits:
     (a) five hundred forty-three billions seven hundred ninety-two millions two hundred seven thousands eight hundreds and twenty-nine;

(b) thirty-four billions seventy-one millions one thousand five hundreds and forty-three;

(c) forty-five billions three hundreds thirty-three;

(d) one billion seven thousands.

25. What does the digit 5 signify in the writing of each of the following numbers?

(a) 253

(b) 576

(c) 15,527

(d) 500,125

(e) 55,550

26. Write all three-digit natural numbers in which only the digits 5 and/or 9 are used.

27. What are the greatest and smallest three-digit natural numbers that may be written, using only the digits 9, 7, and 3?

28. Write all three-digit numbers, using only the digits 3, 1, and 5. Find the sum of these numbers.

29. Pupils X, Y, and Z are waiting to see the Principal of their school. Pupil X would require 7 minutes to talk with the Principal. 5 minutes and 9 minutes are required for pupils Y and Z, respectively. How their seeing the principal may be arranged so that the pupils wait as less time as possible.

30. Solve the following equations:

(a) $2{,}586 + x = 7{,}819$;

(b) $n + 568 = 2{,}222$;

(c) $2{,}845 + (a + 905) = 5{,}959$;

(d) $745 + (r - 1{,}910) = 3{,}562$.

31[*]. The speed of a train is 60 mph, and the speed of a bicyclist is five times as less. How much time is required for the bicyclist to cover the distance which the train traveled in four hours?

32. Write all one-digit to seven-digit numbers, using only the digit 9. What is the place value of the digit 9 in each number?

33. Which number will it be if we add together five tens of hundreds, seven tens, nine units, six hundreds of thousands, and three hundreds of millions?

34. What number should be deducted from 5,943 so that the difference can be 2,525?

35. Read the approximate equalities and tell to which *place value* rank a number has been rounded off:
    (a) $52{,}436 \approx 52{,}400$;
    (b) $74{,}658 \approx 74{,}660$;
    (c) $65{,}917 \approx 70{,}000$;
    (d) $805{,}359 \approx 800{,}000$.

36. Round of the number 864,097:
    (a) to the nearest tens;
    (b) to the nearest hundreds;
    (c) to the nearest thousands;
    (d) to the nearest tens of thousand.

37. Solve the following problems. Round off the answers in the first column to the nearest hundred, and the answers in the second column – to the nearest thousand:

First Column
(a) $73{,}815 \div 3$;
(b) $7{,}680 \cdot 2$.
  Second Column
  (c) $65{,}497 + 8{,}561$;
  (d) $596{,}794 - 104$.

38. One-tenth of the length of the Sanquin River is 56,432 miles. Round off this number to the nearest tens of thousands.

39. The mass of three bags of eddoes is 5 kg 584 g. Round off this quantity to the nearest kilograms.

40. Write the following:
    (a) in kilometers, having rounded off each at first to the nearest millions:
        (i) 24,680,135 meters;
        (ii) 4,128,156 meters;
        (iii) 3,690,258 meters;
        (iv) 9,956,009 meters;
    (b) in dollars, having rounded off each at first to the nearest hundreds:
        (i) 5,431 cents;
        (ii) 845 cents ;
        (iii) 2,523 cents;
        (iv) 284 cents.

41. Examine Fig. 1.12 below. How many of the following geometric figures are represented in it: *points; broken lines; line segments; rays; straight lines;* and *curves*? Name the figures represented.

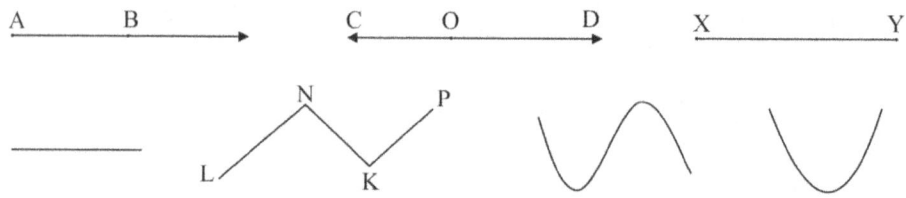

Fig. 1.12. Some geometric figures.

42. Make two points (P and K) and take a straight line L through them.

43. Illustrate two line segments MN and MC (not necessarily drawn to scale) having the lengths of 5 cm and 3 cm, respectively, and find their sum and difference.

44. Draw a line segment CD = 12 cm. Mark on it point P anywhere between points C and D so that CP = 7.25 cm. What is the length of PD?

45. Draw line segments (not necessarily drawn to scale) OP = 40 mm, PT = 50 mm, and TR = 3 cm, such that points O, P, and T, and likewise points P, T, and R do not lie on the same straight line. Determine the lengths of the broken line OPTR so formed.

46. The distance from City X to City Y is 560 miles; and from City X to City Z, the distance is greater by 85 miles. What is the distance from City Y to City Z through City X?

47*. The distance between two points M and N on a straight line is 2 inches. Mark two points on this straight line. The first point should be to the left of point M, and the second point should be to the right of point N. At what distances from M and N, respectively, should the required points be marked on the straight line so that the sum of the distances from each point to points M and N (including the segment MN) is equal to 4 inches. What is the distance between the first point and M? What is the distance between the second point and N? Show the drawing to your answer.

48. Three points are given from one side of a straight line L, and four points from the other side. All of the points are ends of line segments passing through the given straight line L. Draw the figure and tell the number of line segments.

49. Consider the ray in Fig.1.13. Numbers 3 and 5 are marked on it. You are required to mark numbers 1, 2, 4, 6, and 7. Read the name of each ray represented in the figure.

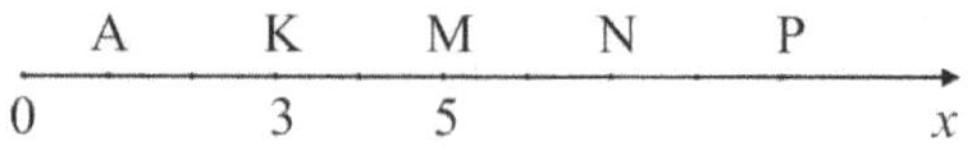

Fig.1.13. A ray for the solution to problem number 49.

50. The length of a unit segment on a number line is 3 cm. Point C on the number line corresponds to the number 5, and point D to the number 17. What is the total length in centimeters of the line segment CD, considering the number of unit segments between points C and D?

51. The height of a tower is 63 ft, and the height of a tree is 37 ft. By how many feet must the tree grow so that it would be 12 ft higher than the tower?

52. Supply the sign " > " or " < " in stead of the asterisks *** , and read the indicated inequalities.
    (a) 785,632 *** 785,652;
    (b) 8,185,183 *** 8,185,103;
    (c) 5,723,141 *** 5,723,041;
    (d) 10,000 *** 9,999.

53. Calculate the value of each expression below and, in stead of the asterisks ***, supply the appropriate sign " > " , " < " , or " = ".
    (a) $503 + 6,450 - 1,140$ *** $4,000 + 4,510 - 549$;
    (b) $6,013 + 154 \cdot 12$ *** $769 \cdot 7 - 85 \cdot 15$;
    (c) $29,000 \cdot 10$ *** $2,900 \cdot 100$.

54. Which is greater?
    (a) 45 kg or 445 g ;
    (b) 12 km or 1200 m;
    (c) 3 km 72 cm or 372 cm;
    (d) 5 kg or 5200 g ?

55. Compare the expressions: $910,799,600 \div 40 - 65$ and $35,286 \cdot 1,574 - 109,845$. Which is greater?

56. Write the greatest and the least sixth-digit numbers consisting of two fours, three fives, and a one.

57. It is possible to travel by bus from City X to City Y with a transit in either City Z or in City P. Which way is more advantageous (in terms of the shorter waiting time), if it is known that from City X to City Z it takes 2 hours 15 minutes, and it is necessary to wait 25 minutes for the next connecting bus; and that it take 35 minutes from City Z to City Y? From City X to City P it takes two hours by bus, and it is necessary to wait 15 minutes in City P for the next connecting bus which takes 55 minutes to arrive in City Y.

Chapter One: Exercises to Rate your Ability

1. Find the result of the indicated operations:
    (a) $7,513 + 20,456 + 890,143$;
    (b) $112,347 - 99,889$;
    (c) $765,432 \cdot 123$;
    (d) $450,132 \div 15$.

2. What is the value of 5 raised to the second power?

3. What is the value of 4^3?

4. What is the cube root of 64?

5. What is the value of the expression $\sqrt{1024}$?

6. What is n, if the value of n raised to the 3rd power is 27?

7. Determine the value of the given expression, observing the order of operations:

    $(6 + 9) \cdot 7 - 4 + 9 \div 3$.

8. Perform the indicated operations, observing the order of operations:
$5 \cdot (6 + 8 \div (4 \cdot 2 - 15)) + 10$.

9. Determine which of the following numbers can be divided by 6: (a) 3,694; (b) 2,578; (c) 2,406.

10. Write the following number in digits: five hundred twenty-six billions six hundred seventeen millions eight hundred fifty-three thousands four hundreds seventy-one.

11. Write the following numbers in words:
(a) 35,641,278; (b) 936,000,547,005

12. Compare the following numbers and write the answer with the help of the signs "<" or ">":
(a)8,006,907 and 8,006,807;
(b)   5,040,302 and 5,040,203.

13. Round off the numbers first to the nearest tens, and then to the nearest hundreds, and then to the nearest thousands:
(a)78,654,123
(b)   23,145,678.

14. Mark on a number line the following points: B (2); C (4); N (7); K (10); and S (12).

15. The total number of 5th and 6th grade pupils in a certain school is 755. Which grade has more pupils and by how much, if it is known that there are 371 pupils in the 5th grade?

16. Write the smallest and the greatest five-digit numbers using the digits 5, 0, 2, and 3.

17. Given the sequence of numbers: 3,003; 3,006; 3,010; 3,020; 3,023; continue the sequence to at least ten numbers.

18.   Between which adjacent natural numbers is found each of the following numbers: 12,003; 503,104; 678; 999?

19.   The digit 7 has been added from the left to a three-digit number. How will this number change? Give an example.

CHAPTER TWO: ADDITIONS AND SUBTRACTIONS

2.1   Properties of Addition

Kofi and Kojo had returned home from finishing. Kofi caught 15 fishes, and Kojo caught 23. When their mother asked them how many fishes they had caught together, the children answered that in order to find out this, she should add the number 15 to the number 23. But their mother preferred to do otherwise by adding the number 23 to the number 15. Will the answer change or not, if at first she takes the 23 fishes caught by Kojo and add them to the 15 fishes of Kofi? We can see that 15 + 23 = 38, and 23 + 15 = 38.

The commutative property (or commutative law) of addition states that the result of an addition operation is the same, irrespective of the order of the addends. In other words, regardless of the arrangement or re-arrangement of the addends in an addition operation, the sum always remains the same. In general, if we denote the addends by the letters x and y, then the following equality is true:

$x + y = y + x$

This means that the result of addition does not depend on the sequence in which the addends are written. For example, 173 + 305 = 305 + 173. Irrespective of the values of the addends, the sum does not change if these addends exchange places.

Let us suppose that there are three buses of tourists on an excursion. In the first bus there are 18 tourists, 15 in the second, and 25 in the third. It is necessary to find out how many tourists in all that are on this excursion. It is possible to find at first the number of tourists in the first and second buses: 18 + 15 = 33. After that we add to this sum the number of tourists in the third bus: 33 + 25 = 58.

It is possible to do otherwise. We first find how many tourists in the second and third buses put together: 15 + 25 = 40; and after that we add to this sum the number of tourist in the first bus: 40 + 18 = 58. In such a way, we have: $(18 + 15) + 25 = 18 + (15 + 25)$. It is easy to observe this property, adding three or more addends. In this lies the gist of the associative property of addition.

> *The associative property of addition states that the result of an addition operation is independent of the grouping of numbers or terms within a given set of addends. In other words, if a third number is added to the sum of two numbers, then the result will be equal to the sum of the first number and the result of adding the second and the third numbers. In general, given any values of the letters x, y, z, the following equality is true:* $(x + y) + z = x + (y + z)$

From these two properties it is possible to re-arrange and combine addends in any order or grouping; for example: $425 + 123 + 515 + 798 = (425 + 515) + (123 + 798) = 940 + 921 = 1861$. This example indicates that the following conclusion is true: *the sum of two natural numbers is always greater than each addend, that is $x + y > x$ and $x + y > y$.*

> There exists only one number addition of which to any natural gives the result of that very same number. Hence this additive property: *if one of two addends is equal to zero, then their sum is equal to the second non-zero addend.* This means that given any values of x, the following equality is true:
> $x + 0 = x; 0 + x = x.$

2.2 Properties of Subtraction

A group of 15 tourists was waiting on a standby to enter into a museum when 6 of them left the group to a nearby shop to buy some things. How many tourists remained in the group? This problem can be solved by subtraction: $15 - 6 = 9$.

If the 6 tourists returned to the group from the shop then there would again be 15 tourists. Therefore, it can be said that the operation of subtraction is reverse or opposite to that of addition; that is, subtraction is an operation in which by the sum (15) of two addends and one of the addends (6) the second addend (9) can be determined. If we examine the natural number line, then it is not difficult to be convinced about the following observation: *to subtract a natural number (6, for example) from 15 means to find on the natural number line such a number after which 15 stands in the sixth place.* See Fig.2.1.

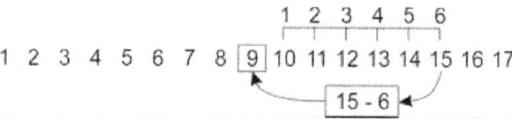

Fig. 2.1. A subtraction operation on the natural number line.

For natural numbers, subtraction is possible only (and only) when the minuend is greater than or equal to the subtrahend. Below we examine different cases connected with *subtraction of a sum from a number; subtraction of a number from a sum; addition of a number to a difference; and subtraction of a difference from a number:*

$x - (y + z)$	- subtraction of a sum from a number;
$(x + y) - z$	- subtraction of a number from a sum;
$x + (y - z)$	- addition of a number to a difference;
$x - (y - z)$	- subtraction of a difference from a number.

In each of the preceding cases, it is possible to carry out a subtraction operation by two methods. The first method lies in the application of the rule of the order of performing arithmetic operations involving expressions with brackets. The essence of the second method may be established in the solutions to the following problems.

Problem 1. A market woman carried 280 kg of rice to the Rally Time Market to sell. In the morning before 12 pm she sold 120 kg of rice.

After 5 pm just before going home she sold additional 95 kg of rice. How many kg of the rice remained?

Solution.
First method: $280 - (120 + 95) = 280 - 215 = 65$ (kg);
Second method: $280 - 120 - 95 = 65$ (kg).
We can therefore draw a conclusion that *in order to deduct a sum of two numbers from a number, it is necessary and enough to consecutively deduct each addend separately:280 − (120 + 95) = 280 − 120 − 95.*

Problem 2. In a school canteen there were 28 kg of corn meal in one bag, and 25 kg in another bag. During lunch time 23 kg of the corn meal was used to bake corn bread. How many kilograms of the corn meal remained in the canteen?

Solution.
First method: $(28 + 25) - 23 = 30$ (kg);
Second method: $(28 - 23) + 25 = 30$ (kg);
Third method: $28 + (25 - 23) = 30$ (kg).
It is easy to see that (28 + 25) − 23 = (28 − 23) + 25 =28 + (25 −23).

Problem 3. Mulbah and his small brother Jallah went to the beach to swim. Mulbah climbed a coconut tree and picked 9 young coconuts and 12 full grown ones. The two brothers gave 5 of the full grown coconuts to their friends. How many coconuts remained with Mulbah and Jallah?

Solution.
First method: $9 + (12 - 5) = 16$ (coconuts);
Second method: $(9 + 12) - 5 = 16$ (coconuts).
It can be seen that 9 + (12 − 5) = (9 + 12) − 5.

Problem 4. Nagbe bought 17 pencils and later gave 6 of them to his sister. The next day he bought additional 3 pencils. How many pencils remained in his possession?

<u>Solution.</u>
First method: $17 - (6 - 3) = 14$ (pencils);
Second method: $(17 - 6) + 3 = 14$ (pencils).
It is clear that $17 - (6 - 3) = (17 - 6) + 3$.

We can extend the conclusion from the first problem to say that it is possible to subtract in parts, for example: $65 - 38 = 65 - (18 + 20) = (65 - 18) - 20 = 47 - 20 = 27$.

Also the solutions to problems 2, 3, and 4, with particular emphasis on the second methods, reveal the application of certain properties that are characteristic of subtraction operation. In each solution it was clearly seen that the result was the same regardless of the method used. Consequently, just as in addition, *the associative property is also applicable in subtraction.* Examples of the *associative property of subtraction* are the following:

    1. $(28 + 25) - 23 = 28 + (25 - 23)$;
    2. $9 + (12 - 5) = (9 + 12) - 5$;
    3. $17 - (6 - 3) = (17 - 6) + 3$.

2.3   Addition and Subtraction Tables

We can represent addition of numbers in the form of a table sometimes called an *addition table*. See Table 2.1.

+	1	2	3	4	5	6	7	8	9	10	11	12	13
1	2	3	4	5	6	7	8	9	10	11	12	13	14
2	3	4	5	6	7	8	9	10	11	12	13	14	15
3	4	5	6	7	8	9	10	11	12	13	14	15	16
4	5	6	7	8	9	10	11	12	13	14	15	16	17
5	6	7	8	9	10	11	12	13	14	15	16	17	18
6	7	8	9	10	11	12	13	14	15	16	17	18	19
7	8	9	**10**	11	12	13	14	15	16	1	18	19	20
8	9	10	11	12	13	14	15	16	17	18	19	20	21
9	10	11	12	13	14	15	16	17	18	19	20	21	22
10	11	12	13	14	15	16	17	18	19	20	21	22	23
11	12	13	14	15	16	17	18	19	20	21	22	23	24
12	13	14	15	16	17	18	19	20	21	22	23	24	25
13	14	15	16	17	18	19	20	21	22	23	24	25	26

Table 2.1. An addition table.

In order to find the sum of two numbers in an addition table, we proceed like this: we take one addend (3, for example) in the upper first horizontal row and a second addend (7, for example) in the first left column of the table. Then the sum will be in the square at the intersection of the corresponding column under the digit 3 and the row which begins with the digit 7; that is, the sum will be **10**.

Similar case of subtraction which we can make up in the form of an analogical table is called a *subtraction table*. Results of subtraction table are determined on the basis of addition or by subtraction of a number in parts.

Calculation by the first method: Suppose it was required to find the difference of $15 - 6$; 15 is $6 + 9$. That is, if we deduct 6 from 15, we are remained with 9.

Calculation by the second method:
$$15 - 6 = 15 - (3 + 3) = (15 - 3) - 3 = 12 - 3 = 9.$$

2.4   Oral Addition and Subtraction Operations

Addition and subtraction are carried out either orally or in writing by defined rules using the properties of these operations. We can perform written calculations in all cases. However, on this level, oral calculations as a rule are performed in the limits of two-digit or three-digit numbers and by carrying out operations over them in that range. The basic method of oral calculations is an approach to the addition and subtraction of numbers involving the use of place values of digits. The performance of operations starts from the highest place value ranks. Let's examine the writing below and explain the approach to such oral calculations:

(a) $34 + 61 = (30 + 60) + (4 + 1) = 90 + 5 = 95$;
(b) $43 + 27 = (40 + 20) + (3 + 7) = 60 + 10 = 70$;
(c) $59 - 18 = (50 - 10) + (9 - 8) = 40 + 1 = 41$;
(d) $98 - 39 = (80 - 30) + (18 - 9) = 50 + 9 = 59$.

In the addition or subtraction of two-digit numbers with conversion through a group of tens it is possible to apply the method of successive or sequential performance of operations. Let's look at the following examples:

(a) $23 + 57 = (23 + 50) + 7 = 73 + 7 = 80$;
(b) $84 - 65 = (84 - 60) - 5 = 24 - 5 = 19$.

Usually, when three-digit numbers are involved, the method of rounding off numbers is widely applied in oral calculations, as in the following examples:

(a) $543 + 302 = 543 + 300 - 2 = 841$;
(b) $987 - 570 = 987 - 600 + 30 = 417$.

It would be useful to try to explain how the rounding-off operations were carried out in the preceding examples.

2.5   Written Addition and Subtraction Operations

2.5.1   Written Addition Operations with Natural Numbers

In written addition numbers are written so that digits of the same place value ranks are in one vertical column. In other words, *units* are placed under *units, tens* under *tens, hundreds* under *hundreds*, and so on. An addition sign is placed to the left of the addends, usually before the last addend at the bottom of the column. Also a horizontal line is drawn below this last addend under which the sum is written.

As a rule, the process of addition is reduced to the addition of one-digit or simple numbers of each column. People begin to add from the lowest place value digits, i.e. from the *unit* digits of each separate addend. If the sum of the addends of a column is a two-digit number, then its *units* digit is written in that very column below the horizontal line, while the number of tens (or the *tens digit*) is *carried over* or added to the numbers of the next place value rank. This process is continued from the first *right* column to the last *left* column. Below we apply the rule of written addition to check the following examples:

(a)   52,610 - Addend
     + 364,495 - Addend
       417,105 - Sum

(b)   32,123 - Addend
         9,476 - Addend
       545,367 - Addend
     +   95,678 - Addend
       682,644 – Sum

2.5.2   Written Subtraction Operations with Natural Numbers

The order of written subtraction is similar to that of addition. The minuend and subtrahend are written so that digits of the same place value ranks are in one vertical column, i.e. *units* are placed under *units, tens* under *tens, hundreds* under *hundreds*, and so on. Just like in addition, the process of subtraction is reduced to the subtraction of simple one-digit numbers of each column. People begin subtraction from the lowest place value digits; that is by subtracting the unit digit of the subtrahend from the unit digit of the minuend in each column.

If a digit in the subtrahend is greater than the corresponding digit in the minuend, then a 1 (which is practically equal to one group of ten due to the base-ten system we are using) is *borrowed or carried over from* the place value rank at its left in the minuend. Also a subtraction sign is placed to the left of the subtrahend. A horizontal line is drawn below the subtrahend, under which the difference is written. Some examples of written subtraction are given below:

(a) 987,654 – Minuend
    – 586,343 – Subtrahend
    401,311 - Difference

(b) 751,436 - Minuend
    – 213, 954 - Subtrahend
    537,482 - Difference

In order to check the correctness of a subtraction operation, the *difference* and the *subtrahend* are added; if the resulting *sum* is equal to the *minuend,* then the subtraction is performed correctly. What other method do you think can be used to check the correctness of a subtraction operation?

2.5.3   Addition of Rational Numbers

So far we have discussed written addition and subtraction operations and given some examples. It is still necessary and essential to give more basic or elementary explanations of these concepts and operations with the help of coordinate straight lines, either on the vertical y-*axis* called *ordinate,* or on the horizontal x-*axis* called *abscissa.* The numbers on the right of the *zero* on the abscissa are *positive numbers*, while those on the left of the *zero* are *negative numbers.* In much the same way, the numbers above the *zero* on the ordinate are *positive numbers*, while those below the *zero* are *negative numbers.*   It should be noted that given any two numbers on the abscissa, the one on the right is always greater; i.e. the numbers on the abscissa increase in counting from left to right. Likewise, given any two numbers on the ordinate, the upper one is always greater, i.e. the numbers on the ordinate increase in counting from down upward. On each coordinate straight line, *zero* is the starting point, or the so-called *point of reference*.

Let us look at a few example problems below:

Problem 1. If the reading on a thermometer was +7 ^{0}C and later the reading had changed by +2 ^{0}C, then how many degrees does this thermometer now show?

Solution: $(+7) + (+2) = +9$ – the present reading on the thermometer is +9 ^{0}C .

Problem 2. If the reading on a thermometer showed − 5 ^{0}C and after 24 hours the reading changed by − 6 ^{0}C, then how many degrees does this thermometer now show?

Solution: $(− 5) + (− 6) = −11$ – the present reading on the thermometer is −11 ^{0}C.

Problem 3. If the reading on a thermometer showed +5 ^{0}C and after a certain interval of time the reading changed by − 10 ^{0}C, then what is the new reading on this thermometer?

Solution: $(+5) + (− 10) = −5$ – the new reading on the thermometer is −5 ^{0}C.

Problem 4. Represent the number +3 as point K on a horizontal coordinate line (*x-axis*). Transfer point K by 6 unit segments to the right to point L. Find the coordinate of point L on the *x-axis*.

Solution: $(+3) + (+6) = +9$, the coordinate of point L on the *x-axis* is +9; it may be written as L(9).

Problem 5.
(a) Transfer point Q (−2) by −3 unit segments to point P on a horizontal coordinate line (*x-axis*). Find the coordinate of point P.

Solution: (−2) + (−3) = −5, the coordinate of point P on the *x-axis* is −5;

Note: *If the number of unit segments by which you are required to transfer a point is minus, then the direction of transfer is to the left on the x-axis. The required coordinate of point P may be written as P(−5).*

(b) Transfer point M (−12) by +5 unit segments to point N on a horizontal coordinate line (*x-axis*). Find the coordinate of point N.

Solution: (−12) + (+5) = −7, the coordinate of point N on the *x-axis* is −7;
Note: *If the number of unit segments by which you are required to transfer a point is plus, then the direction of transfer is to the right on the x-axis.*

Problem 6.
(a) Transfer point R (−5) by +9 units to point G on a horizontal coordinate line (*x-axis*). Find the coordinate of point G.

Solution: (−5) + (+9) = +4, the coordinate of point G on the *x-axis* is +4. See Fig. 2.2 below.

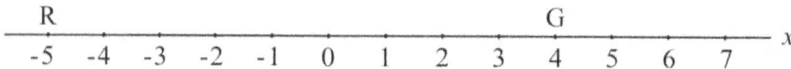

Fig. 2.2. Addition of rational numbers

(b) Transfer point D (+8) by −6 units to point C on a horizontal coordinate line (*x-axis*). Find the coordinate of point C.

Solution: $(+8) + (-6) = +2$, the coordinate of point C on the *x-axis* is $+2$.

From the above operations, it is easy to understand that *the sum of two positive numbers is greater than each of the addends*, as indicated in problems 1 and 4; and *that the sum of two negative addends is less than each of the addends* (problems 2 and 5 (a)). It should be noted that -11 is less than -6; and -6 is less than -5. Likewise -5 is less than -3; and -3 is less than -2. However, *the module (absolute magnitude or absolute value) of the sum of two negative addends is greater than each of the addends.*

Consequently, judging from this context, it is possible to formulate the following rule:

The sum of two numbers with identical signs is a number, having the same sign as the addends; and its module is equal to the sum of the modules of the addends.

There is a different situation regarding the sum of two addends, when one of them is *negative*. Looking at the solutions to problems 3, 5 (*b*), 6 (*a* and *b*), we can formulate the following rule:

The sum of two numbers with different signs is a number, having the same sign as the addend with the greater module; and its module is equal to the difference of the modules of the addends.

Check the truth of the two given rules by means of the example problems above.

2.5.4   Subtraction of Rational Numbers

Let us look at some example problems below:

Problem 1. Find the root of each equation:

(a) $18.5 + b = 22.6$
(b) $6.7 = 1.5 + m$
(c) $12.04 = 3.98 + n$
(d) $21.3 + a = 27.1$

Problem 2. Solve the equations:
   (a) $7.4 - c = 1.9$
   (b) $d - 9 = 11.1$
   (c) $x = 11 - 16$
   (d) $8 - 15 = y$

Problem 3. Solve the equations:
   (a) $b - 17 = -12$;
   (b) $k + 29 = 18$;
   (c) $25 - r = 14.7$;
   (d) $6.5 + w = 2.4$.

Solutions to Problem 3:

(a) $b - 17 = -12$; adding +17 to both sides of the equation will not change the equality;
$b - 17 + (+ 17) = -12 + (+17)$; $((-17 + 17) = 0$; $\rightarrow$ $b + (-17 + 17) = b)$;
$b = -12 + 17$;
$b = 17 - 12$; (commutative property);
$\boldsymbol{b = 5}$ – the root of the equation;

(b) $k + 29 = 18$; adding -29 to both sides of the equation will not change the equality;
$k + 29 + (-29) = 18 + (-29)$; $((+ 29 - 29) = 0$; $\rightarrow k + (29 - 29) = k)$;
$k = 18 - 29$;
$k = -29 + 18$; (commutative property);
$k = -29 + 18$; (Apply the rule about the *sum of two numbers with different signs*);
$\boldsymbol{k = -11}$ (the root of the equation);

(c) $25 - r = 14.7$; (Since the *unknown* or *variable* is negative, we have to multiply every term of the equation on both sides by -1 in order to make it a positive variable; this will not change the equality.);

$r - 25 = -14.7$; adding $+25$ to both sides of the equation will not change the equality;

$r - 25 + (+25) = -14.7 + (+25)$;

$r = -14.7 + 25$;

$r = 25 - 14.7$; (commutative property);

$r = \mathbf{10.3}$ – the root of the equation;

(d) $6.5 + w = 2.4 \rightarrow w + 6.5 = 2.4$; (Commutative property applied to the left side of equation);

$w + 6.5 = 2.4$;

$w + 6.5 + (-6.5) = 2.4 + (-6.5)$; adding -6.5 to both sides of the equation will not change the equality;

$w = 2.4 - 6.5 \rightarrow w = -6.5 + 2.4$; (Apply the rule about the *sum of two numbers with different signs*);

$w = \mathbf{-4.1}$ (the root of the equation).

As we can see from the solutions to problem 3 above, *it is possible to replace a subtraction operation by the addition of an equal number which is opposite in sign to the subtrahend.*

Another example is finding the root of the equation $x = 3 - 10$. Finding the difference of two numbers is reduced to finding the sum of two numbers with different signs:

$x = 3 - 10$;

$x = 3 + (-10)$;

$x = -7$.

In view of the above, we can formulate the following rule:

In order to find the difference between two non-zero numbers, it is necessary and sufficient to add to the minuend a number which is opposite to the subtrahend:

$$m - n = m + (-n).$$

Problem 4. Replace the subtraction of numbers by the addition of rational numbers and perform the following calculations.

(a) $(15.3) - (+6) - (+8) + (-17) - (-9) - (5.2)$;

Solution: $(15.3) - (+6) - (+8) + (-17) - (-9) - (5.2)$;
$= (15.3) + (-6) + (-8) + (-17) + (+9) + (-5.2)$;
$= (15.3) + (+9) + (-6) + (-8) + (-17) + (-5.2)$;
$= (15.3 + 9) + (-6 - 8 - 17 - 5.2)$;
$= 24.3 + (-36.2)$;
$= -11.9$;

(b) $(+3.5) - (+7) - (+1.7) - (-6.6) + (+5.1)$;

(c) $+12 + (-8) + (+6) - (+9) + (-4) - (-2)$;
Solution: For simplifying the writings of numbers, it is possible to omit the addition signs, remembering the sign which stands before each number in a given expression. We can therefore write in a short form the expression $+12 + (-8) + (+6) - (+9) + (-4) - (-2)$ as:

$$12 - 8 + 6 - 9 - 4 + 2$$
$$= 12 + 6 + 2 - 8 - 9 - 4$$
$$= 20 - 21$$
$$= -1$$

2.5.5   More on Powers or Exponents

In section 1.1 of Chapter One of this book we talked a little about *raising a number to a power* and the *extraction of a root*. *Raising a number to a power* and *extracting a root* are other arithmetic operations which are frequently encountered in the study of mathematics. In this section we concentrate only on raising a number or quantity to a power.

Let's consider the following problems:

Problem 1. What is the area of a square whose side is equal to 2.6 inches?

Problem 2. There are five pencil-cases in each of which are five pencils. There are how many pencils in all?

Problem 3. Find the volume of the cube in Fig.3.8 (section 3.11) where n, the edge, is assumed to be equal to 4.5 centimeters.

Solution: As we already know, the volume of a cube is equal to the cube of its edges:

$$V_{cube} = n^3;$$
$$V_{cube} = (4.5)^3;$$
$$V_{cube} = (4.5) \cdot (4.5) \cdot (4.5);$$
$$V_{cube} = 91.125 \text{ cm}^3 \text{ (or } 91.125 \text{ cubic centimeters);}$$

Problem 4. Find the value of each indicated exponent below:

(a) (i) $(3\frac{1}{2})^4$; (ii) $(-1\frac{2}{5})^5$; (iii) $(4\frac{5}{8})^0$; (iv) $(-0.2)^2$;

(v) $-(9)^3$; (vi) $(\frac{3}{4})^{-1}$; (vii) 8^{-1};

(b) (i) $(-5)^{-2}$; (ii) $-(3)^{-3}$; (iii) $-(6)^4$; (iv) $(-6)^4$; (v) $(\frac{3}{4})^{-2}$;

(vi) $(-\frac{3}{4})^{-2}$; (vii) x^{-2};

Solution.

(a) (i) $(3\frac{1}{2})^4 = (\frac{7}{2})^4 = (\frac{7}{2}) \cdot (\frac{7}{2}) \cdot (\frac{7}{2}) \cdot (\frac{7}{2}) = (\frac{7 \cdot 7 \cdot 7 \cdot 7}{2 \cdot 2 \cdot 2 \cdot 2}) = \frac{2401}{16} = 150\frac{1}{16}$;

(iii) $(4\frac{5}{8})^0 = 1$; every number or quantity raised to the power of zero is equal to 1;

(v) $- (9)^3 = - (9 \cdot 9 \cdot 9) = - (729) = - 729;$

(vii) $8^{-1} = \frac{1}{8};$

(b) (ii) $- (3)^{-3} = - (\frac{1}{3^3}) = - (\frac{1}{3 \cdot 3 \cdot 3}) = - (\frac{1}{27}) = - \frac{1}{27};$

(iv) $(-0.2)^2 = (-0.2) \cdot (-0.2) = 0.04;$

(vi) $(-\frac{3}{4})^{-2} = \frac{4 \cdot 4}{3 \cdot 3} = \frac{16}{9} = 1\frac{7}{9};$

(vii) $x^{-2} = \frac{1}{x^2} = \frac{1}{x \cdot x} = \frac{1}{x^2}.$

In the solution to problem 3, the number *4.5* is the *base* and *3* is the *power* or *exponent*. The *power* or *exponent indicates* how many times a number or quantity is multiplied by itself; and the *base* is the number or quantity that is multiplied by itself. **Exponentiation** (or raising to a power) of positive numbers was explained in section 1.1. Therefore, the focus here is to examine the case when the base of the power or exponent is a negative number. Other similar cases are demonstrated in problem 4 as to how to proceed with the solution of a problem when the power or exponent itself is a negative number. As an example, let us consider exponentiation of the number -3:

$(-3)^0 = 1;$

$(-3)^1 = -3;$

$(-3)^2 = (-3) \cdot (-3) = 9;$

$(-3)^3 = (-3) \cdot (-3) \cdot (-3) = -27;$

$(-3)^4 = (-3) \cdot (-3) \cdot (-3) \cdot (-3) = 81;$

$(-3)^5 = (-3) \cdot (-3) \cdot (-3) \cdot (-3) \cdot (-3) = -243;$

$(-3)^6 = (-3) \cdot (-3) \cdot (-3) \cdot (-3) \cdot (-3) \cdot (-3) = 729.$

We can see that in the exponentiation of the number -3, a *positive* resulting number appears in each *even* power, while in an *odd* power there appears a *negative* resulting number. It is obvious that such regular succession of signs results in the exponentiation of any negative number. Consequently, we can formulate the following rules: *An even exponent of a negative number is positive; and an odd exponent of a negative number is negative.*

2.5.6   Coefficients

*A **coefficient** is a numerical factor of a literal expression.* For example, the coefficient of the expression $5y$ is the number 5. The coefficient of the expression $8.3k$ is the number 8.3; the coefficient of the expression xy is 1 since $xy = 1 \cdot xy$; and so on. If a product contains several numerical factors, then in order to find the coefficient it is necessary to multiply them together. For example, the coefficient of the expression $0.5m \, (3n)$ is equal to 1.5 since $0.5m \, (3n) = 1.5mn$.

Let's consider the following problems regarding the operation with coefficients:

Problem. Simplify each of the following expressions:

(a) (i) $4x + 7.4x - 9.5x - 2.1x$;

    (ii) $5m - 6m + 12.7m$;

    (iii) $- 8.5cd + cd + 0.7cd$;

(b) (i) $11k + 5k - 15k$;

    (ii) $-3b - b - 6b$;

(iii) $- 9.8p + 10.1p - 0.8p$

In the problem (a) (i), we can see that the expression $4x + 7.4x - 9.5x - 2.1x$ has the same letter for each term. Thus, we can add them together, i.e. $4x + 7.4x - 9.5x - 2.1x = 11.4x - 11.6x = - 0.2x$.

Addends or *terms* which are different from each other only by their numerical coefficients, but having the same letter or letters, are called *like terms*. Replacement of the sum *5m* − *6m* + *12.7m* by the expression $11.7m$ is called a *reduction of like terms*.

In order to reduce *like terms* it is necessary to add together their coefficients and then multiply the result by the common letter or letters.

2.5.7   Basic Properties of Equation

To begin with, we should consider solving some typical equation problems below:

(a) $y - 3 = 11$;

$\quad\quad y - 3 + 3 = 11 + 3$; adding +3 to both sides of the equation will not upset the equality of the   equation;

$\quad y = 14$ - the solution or root to the equation;

(b) $12x + 15 = 29$;

$12x + 15 - 15 = 29 - 15$; adding −15 to both sides of the equation will not upset the equality;

$\quad 12x = 14$;

$$\frac{12x}{12} = \frac{14}{12};$$ dividing both sides of the equation by 12 will not upset the equality;

$$x = \frac{7}{6} \rightarrow x = 1\frac{1}{6}$$ - the solution or root to the equation;

(c) $9x - 4 = 3x + 6$;

$9x - 3x = 6 + 4$; combining like terms of the equation;

$6x = 10$; reducing like terms of the equation;

$$\frac{6x}{6} = \frac{10}{6};$$ dividing both sides of the equation by 6 will not upset the equality;

$$x = \frac{10}{6} \rightarrow x = \frac{5}{3} \rightarrow x = 1\frac{2}{3}$$ - the solution or root to the equation;

(d) $\frac{3m+1}{5} = -4$;

$(\frac{3m+1}{5}) \cdot 5 = (-4) \cdot 5$; multiplying both sides of the equation by 5 will not upset the equality;

$3m + 1 = -20$;

$3m + 1 - 1 = -20 - 1$; adding -1 to both sides of the equation will not upset the equality;

$3m = -21$;

$$\frac{3m}{3} = -\frac{21}{3};$$ dividing both sides of the equation by 3 will not upset the equality;

$m = -7$; the solution or root to the equation;

From the solutions to the problems above, we can formulate some basic properties of equation:

- *If the same number or quantity is added to both sides of an equation or subtracted from both sides of an equation, then the equality of the equation remains unchanged or unaffected.*
- *If both sides of an equation are multiplied or divided by the same non-zero number or quantity, then the new equation remains the same as the initial equation, i.e. equal to the initial equation.*

2.6   Angles and Polygons

2.6.1.   Angles

If two rays are drawn from a point, then a geometric figure is formed. Such a figure is called an *angle*. The common beginning of the rays forming an angle is called the *vertex* (or *apex*), and the rays themselves are said to be known as the *sides* of the angle. An angle is designated by the symbol "∟" and three capital letters. For example, the angles ∟MPN, ∟ABC, and ∟XYZ have been represented in Fig.2.3 below.

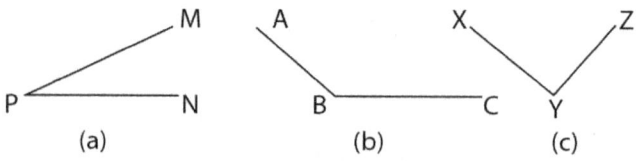

(a)   (b)   (c)

Fig.2.3. Naming of angles: (a) ∠M PN   (b) ∠ABC   (C) ∠XYZ

∠MPN may be read either as *angle MPN* or *angle NPM*. The **middle letter** is the letter designating the **vertex** of the angle. Sometimes an angle is named by a single letter designating its vertex. This means we can still write ∠MPN simply as ∠P.

If we can superimpose two angles one upon the other so that they coincide, then such angles are said to be *equal*. Equal angles have equal measurements. If the sides of an angle are complementary rays of one straight line, then the angle is said to be known as an open or *extended angle*. An open angle is sometimes referred to as a straight angle. Consequently, the terms **open angle, extended angle, straight angle**, or **straight line** may be interchangeably used to refer to an angle measurement of which is equal to 180^0. An open angle ∠ MON is shown in Fig.2.4.

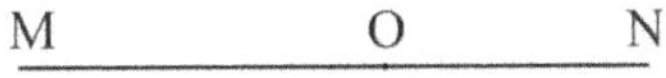

M   O   N

Fig. 2.4. An open angle MON (∠MON) is also a straight line.

If either ray OM or ON of open angle ∠ MON is bent so that its sides (OM and ON) are strictly perpendicular to each other, then the open angle MON is transformed into a different angle referred to as a *right angle*. A **right angle** is an angle measurement of which is equal to 90^0. A right angle is usually designated by the symbol "∟". Examples of right angle are indicated in Fig.2.5. Two straight lines are said to be *perpendicular* if they intersect forming a right angle (s). Straight lines MP and QN in Fig.2.5 are *perpendicular*. The sign designating or indicating that two straight lines are perpendicular is ⊥. In such case, we write "MP⊥QN", which means "line MP is perpendicular to line QN", or "lines MP and QN are perpendicular".

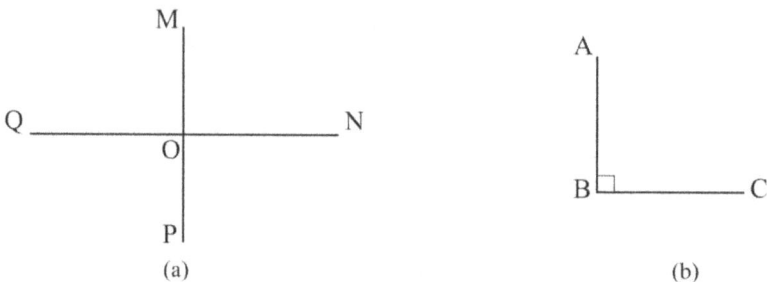

(a)   (b)

Fig.2.5. Examples of *right angles* are $\llcorner$ MON, $\llcorner$ PON, $\llcorner$ MOQ, POQ, and $\llcorner$ ABC.

Angles are measured in **degrees**. *One degree* is equal to one ninetieth ($\frac{1}{90}$) part of a right angle, or $\frac{1}{360}$ of the degree measure of a *circle*. As stated above, a right angle is equal to ninety degrees (90^0). A **circle** is equal to three hundred and sixty degrees (360^0). A **straight line** (which is also considered as a **straight angle**) is equal to 180^0.

Angles are measured with the help of a special device called *protractor*. A protractor is shown in Fig.2.6. It consists of a *ruler* and a **semi-circle** measuring 180^0; the center of the *semi-circle* is located in the middle of the ruler and marked by a small line. The divisions of the scale of a protractor are drawn or plotted on the semi-circle. Each division is equal to one degree (1^0).

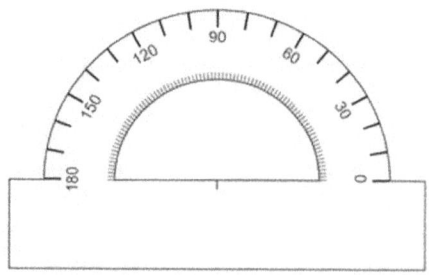

Fig.2.6. A protractor.

In order to measure an angle, we lay on the angle a protractor in such a way that the vertex of the angle coincides with the center of the protractor. One of the sides of the angle passes through the *beginning of reading* (designated by the number zero) on the scale of the protractor. The other side of the angle goes through the graduations on the scale of the protractor and shows the degree measure of the angle. Fig.2.7 (a) shows the measurement of angle ∠ LOP, which is equal to 45^0; and the angle ∠ COD in Fig.2.7 (b) represents a degree measure of 120^0.

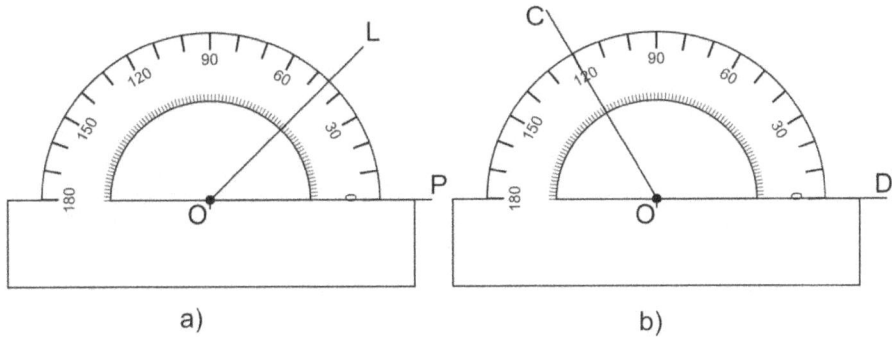

a)   b)

Fig.2.7. Measurement of angles using a protractor. (a) ∠ LOP is equal to 45^0; (b) ∠ COD is equal to 120^0.

We can construct angles with the help of a protractor and also divide them into types. If the degree measure of an angle is less than 90^0, then such angle is called an ***acute angle***. An angle, degree measure of which is more than 90^0, but less than 180^0, is known as an ***obtuse angle***. For example, angle ∠ ABC in Fig.2.3 (b) and angle ∠ XYZ in Fig.2.3 (C) are both obtuse angles. Angle ∠ MPN in Fig.2.3 (a) is an example of acute angle. Examples of a right angle are given in Fig.2.5. Measure the afore-said angles and be convinced that each angle corresponds to its type as defined above.

Two angles are said to be ***complementary*** if the sum of their degree measures is equal to 90^0. A ***supplementary angle*** is either of two angles whose sum is 180^0.

2.6.2 Polygons

Let us consider two different types of broken lines as represented by (a) and (b), respectively, in Fig.2.8. They are different because points K and P of the broken line KLMNP can be joined by yet another line segment; but the end and beginning of the broken line ORSWT are joined.

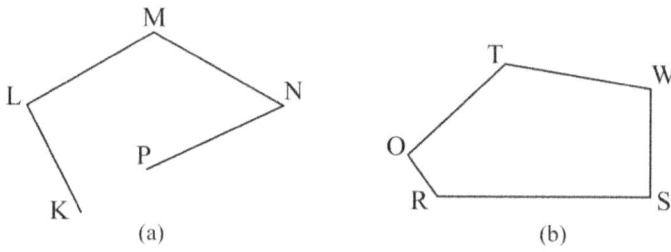

(a)   (b)

Fig.2.8. Two different kinds of broken lines.

A broken line such as ORSWT in which the end and the beginning are joined or coincide is said to be *closed* or *exclusive*. It divides the plane into two parts: an ***internal domain*** (located within the broken line) and an ***external domain*** (located outside of the broken line). A closed broken line forms a many-sided geometric figure called a *polygon*. A ***polygon*** is a geometric figure formed by closed broken lines with its internal domain. The broken line itself is said to be the *boundary* or border of the polygon. Examples of a polygon are represented in Fig.2.9.

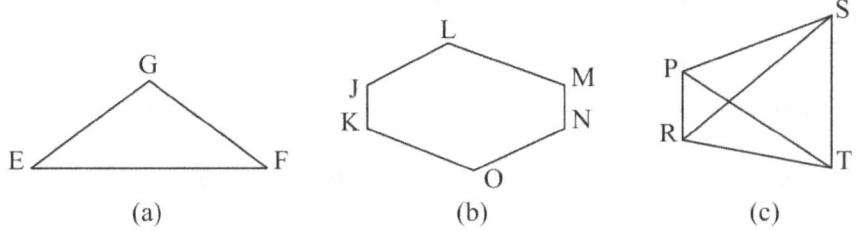

Fig.2.9. Examples of polygons: (a) triangle; (b) hexagon; (c) quadrangle (may also be called *quadrilateral* or *tetragon*).

Separate line segments of the broken line are known as the **sides** of the polygon. Vertices of the broken line are also vertices of the polygon. Vertices of a polygon are designated by capital letters. The sum of the lengths of the separate line segments which are links of the closed broken line is known as the **perimeter** of the polygon.

A line segment connecting two non-adjacent vertices of a polygon is called a **diagonal**. In Fig.2.9 (c), line segments PT and RS are diagonals of the polygon RPST.

The names of polygons are associated with the number of angles they contain. **Triangle, quadrangle, pentagon, hexagon, heptagon, octagon, nonagon** and **decagon** are examples of the names of polygons based on the number of their sides and angles.

2.6.3   Classification of Triangles

The allocation of figures and objects according to definite properties and signs into separate groups or classes is called *classification*. Table 2.2 below represents a classification of triangle on the basis of their angles and sides.
The smallest number of sides which a closed broken line can have is three. It forms the simplest polygon which is a triangle. A triangle may be designated by the symbol Δ. For example, triangle ABC may be represented as "Δ ABC". Depending on the value or degree

measure of the angles, triangles may be distinguished or classified as the following:

(1) **acute triangle** is one in which all the angles are acute angles;

(2) **right- angled triangle** (also called **right triangle**) is a triangle in which there is a right angle;

(3) **obtuse triangle** is one containing an obtuse angle.

Fig.2.10 shows these three kinds of triangles. The sides of a triangle forming a right angle are called **cathetus**. The side lying opposite the right angle in a right triangle is called **hypotenuse**, indicated in Fig.2.10 (a).

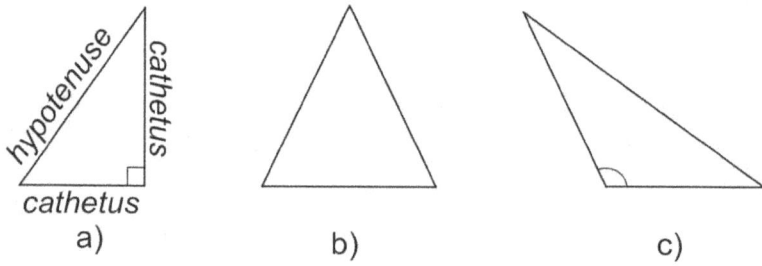

cathetus

a)   b)   c)

Fig.2.10. Classification of triangles. (a) Right angle triangle; (b) Acute triangle; (c) Obtuse triangle.

Sometimes the vertical cathetus and the hypotenuse are referred to as the **legs of a triangle**. The sum of the degree measures of all three angles in any triangle is 180^0. A triangle can have only one right angle or only one obtuse angle.

Depending on the length of the sides, triangles are divided into *scalene triangles* and *isosceles triangles*. **Scalene triangles** are those in which the lengths of all the sides are different. **Isosceles triangles** are triangles in which at least two sides are equal to each other. The two equal sides in an isosceles triangle are called its **lateral sides**.

There are some isosceles triangles in which the lengths of all the three sides are equal. Such triangles are called ***equilateral triangles.***

By sides / By angles	Scalene triangles	Isosceles triangles	
		Non-equilateral	Equilateral
Acute triangles			
Right angle triangles			
Obtuse triangles			

Table 2.2. Classification of triangles.

Consequently, to summarize, we can classify triangles on the basis of their angles into *acute triangle, right triangle,* and *obtuse triangle;* and, on the basis of the lengths of their sides, into *scalene triangle, isosceles triangle,* and *equilateral triangle.*

Another polygon well-known is a *quadrangle.* A **quadrangle** is a four-sided geometric figure. A quadrangle of a general kind may have all of its sides or all of its angles different. However, there are quadrangles in which two, three, or the four sides are equal to one another. Some of them have different names. For example, *rectangle* and *square* are specific kinds of quadrangles. A **rectangle** is a quadrangle in which all of its angles are right angles; and its opposite sides are equal. A **square** is a rectangle in which all of its sides are equal. Fig.2.11 (a) shows a quadrangle of a general type, Fig.2.11 (b) a rectangle, and Fig.2.11 (c) a square. The *perimeter of a quadrangle* is the sum of the lengths of its sides. *In a given quadrangle, only two diagonals may be drawn.*

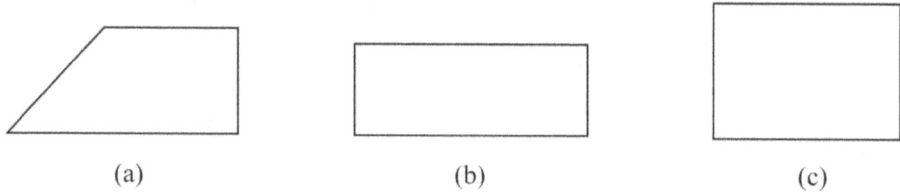

(a)   (b)   (c)

Fig.2.11. Kinds of quadrangle: (a) A quadrangle of a general kind; (b) A rectangle; (c) A square.

2.7   Physical Quantities

In daily life situations, we encounter experience with such physical quantities as *time, distance, velocity or speed, length, force, mass, weight, moment, cost, area, temperature, capacity, density*, and so on. The list is almost inexhaustible. From time to time, you will have to deal with all of these quantities in the course of your study of mathematics and nearly all other subjects.

Physical quantities may be either dependent or independent on one another. For example: the quantity of force which a body in motion acquires depends on its mass and the amount of acceleration; the distance traveled by a body depends on its speed of motion and the time interval in which the motion takes place; the cost of a commodity depends on its price and quantity; and the quantity of heat which a body acquires or loses (radiates) depends on its mass and the change in its temperature.

Let us consider the correlation or dependence between the sides (length and width) of a rectangle and its perimeter. If the length and width of a rectangle are equal to 3 cm and 2 cm, respectively, then its perimeter is equal to 10 cm. We increase each side of the given rectangle by 3 times, and determine the perimeter of the new rectangle. We can compare the perimeters of the two rectangles. If we decrease the lengths of the sides of the second rectangle by 2 times, then we can also calculate its perimeter and compare with the perimeter of the preceding rectangle. In this exercise, we can observe that the perimeter of the third rectangle is two times smaller than the second; and that the perimeter of the second rectangle is three times as great as the first.

We have seen that *increasing (or decreasing) the sides of a rectangle by three (or two) times likewise increases (or decreases) its perimeter by three (or two) times.* Such dependence between quantities is said to be *directly proportional,* and the quantities involved are known to be directly proportional. Therefore, the following statement is true: *two quantities are said to be directly proportional if as a result of increasing (or decreasing) one of them by certain times the value of the other quantity likewise increases (or decreases) by as much times.*

For example, the following quantities are found to be in directly proportional dependence: *heat and temperature; speed and distance; quantity of good and its cost; etc.* The number showing how many times one quantity is more or less than another is called a *ratio.* The preceding exercise has been illustrated in Table 2.3 , where it is easy to see that the sides of a rectangle and its perimeter are directly proportional. Here the ratio is equal to 2. That is, the perimeter of the rectangle is greater than the sum of its sides by two times:

$$\frac{10}{(2+3)} = \frac{30}{(9+6)} = \frac{15}{(4.5+3)} = 2.$$

Rectangle	Length	Width	Perimeter	Remark
	3 cm	2 cm	P = 2 (3+ 2) = 10 cm	Original perimeter of rectangle
	Increased by 3 times: 3 x 3 = 9 cm	Increased by 3 times: 3 x 2 = 6 cm	P = 2 (9 + 6) = 30 cm	30 ÷ 10 = 3; Perimeter increased by 3 times

	Decreased by 2 times: $9 \div 2$ = 4.5 cm	Decreased by 2 times: $6 \div 2 = 3$ cm	P = 2 (4.5 + 3) = **15 cm**	$30 \div 15 = 2$; Perimeter decreased by 2 times

Table 2.3. Direct proportionality between the sides and perimeter of a rectangle.

If the length and width of a rectangle are *increased* or *reduced* by the same number of times, then the resulting perimeter of that rectangle is *increased* or *reduced* by the same number of times.

To be sure that the dependence between two quantities is directly proportional, one has to find the quotient from dividing some pairs of numbers of corresponding values of these quantities. If the resulting quotients or ratios are equal, then the dependence would be a direct proportionality.

Let us consider another example. The distance between Monrovia and Tapita City, for instance, is about 160 miles. A bicyclist who is traveling with a speed of 16 mph would cover this distance within 10 hours. But a motorist traveling with a speed of 32 mph (i.e. twice as fast as the bicyclist) could travel the distance within 5 hours. For a pedestrian whose speed is about 4 mph, a time interval of 40 hours is required in order to overcome the same distance. This is to say that it would require the pedestrian 8 times more and 4 times more of traveling time than the motorist and bicyclist, respectively. In a nutshell, if the distance is unchanged, an increase in the speed of motion will result in a decrease in the time interval of motion.

We can see that *increasing (or decreasing) the speed of motion by a certain value, the interval in which it is possible to overcome the same distance also decreases (or increases) by as much value.* Such dependence between quantities is said to be *inversely proportional,* and the quantities compared are considered to be inversely proportional. Consequently, the following statement is true: *two quantities are said to be inversely proportional if as a result of increasing (or decreasing) the value of one of them by certain times, the second quantity is decreased (or increased) by as much times.*

The following quantities, as a rule, are found in inverse proportionality: the *wholesale price* and *quantity of goods* that can be bought; the *number of workers* and the *time* required to do a given piece of work; and so on. You should think of other physical quantities that are inversely proportional.

It should be noted that different people in different countries are used to measuring the values of quantities in different ways. For example, *length* was measured in *sazhens* (a Russian obsolete length measure equivalent to 2.13 meter) and *cubits. Length* is also measured in *feet, miles, centimeters,* and *kilometers* in different countries. Measures of *weight* were likewise different. The two main known systems of measurements are the *British System* and the *Metric System.* This created a definite inconvenience of interaction in international trade and commerce. Most countries today have eventually converted to the use of the Metric System. It is based on the decimal system; and it lends itself to easier conversion among quantities.

Units of length and mass measurements in the Metric System are shown in Table 2.4. Comparing the indicated entries, it is obvious that relation between units of measurements is 10, 100, 1000, etc.

Measures	Length	Mass
I.	10 mm = 1 cm 10 cm = 1 dm 10 dm = 1 m	1000 mg = 1 g 1000 g = 1 kg 1000 kg = 1 ton 1,000,000 g = 1 ton
II.	100 mm = 1 dm 100 cm = 1 m	10 centner = 1 ton
III.	1000 mm = 1 m 1000 m = 1 km 1,000,000 m = 1 km	1 centner = 100 kg

Table 2.4. Basic units of length and mass measures

The *Metric system* is so-called because the most widely used standard of measurement is one in which the basic unit of length is the *meter*. In the Metric System other measures of length are derived with the help of prefixes which originate from Latin and Greek numerals. *Latin prefixes* such as *deci, centi,* and *milli* are used to designate units less than the basic quantity being measured, whereas *Greek prefixes* such as *deca, hecto,* and *kilo* designate units more than the basic quantity.

Units have special names. Table 2.5 gives a more exact explanation of the Latin and Greek prefixes used in the designation of units of length applicable in the Metric System of measurement.

Units of Length	Meter	Decimeter	Centimeter	Millimeter
Kilometer	1,000	10,000	100,000	1,000,000
Meter	1	10	100	1,000
Decimeter	$\frac{1}{10}$	1	10	100
Centimeter	$\frac{1}{100}$	$\frac{1}{10}$	1	10
Millimeter	$\frac{1}{1000}$	$\frac{1}{100}$	$\frac{1}{10}$	1

Table 2.5. Relationship between length units in the Metric System

2.8 Expression, Equations, and Inequalities

2.8.1 Equations and Equalities

From beginning classes, we are aware that mathematical symbols include the following: digits (0, 1, 2 , 3, ..., 9); letters of the Latin alphabet (A, a, B, b, C, c, ..., Z, z); signs of arithmetic operations (+, −, x or ·, ÷); signs of relations (=, <, >); and parenthesis ( ).

Symbols that are made up of numbers, letters, and joined by signs of mathematical operations are called *expressions*. Expressions which consist only of numbers are called *numerical expressions*. Those containing at least a letter are called *letter expressions*. For example, 4 + 6, 8 ÷ 2 +5, 53 − 21 ÷ 3 are all numerical expressions; $3x + 2$, $5 - y$, and $(7 - a) ÷ 3$ are classified as letter expressions. The properties of addition and subtraction previously discussed were written with the help of expressions.

The area of a rectangle may be calculated as equality, using a letter expression:

$$A = l \cdot w,$$

where A stands for the area, l − the length, and w − the width of the rectangle. Such equalities are known *formulas* or *formulae*. We can remember that with the help of formulas we have already written the properties of certain mathematical operations.

It is obvious that the value of a letter expression or a formula depends on the values of the letters which enter into it. If in the expression *7x + 5* instead of the letter x we assign a value, say *x = 3,* then we have the numerical expression *7 · 3 + 5,* the value of which is equal to 26. If *x = 4,* then the value is equal to 33; and if *x = 0,* then the value is equal to 5.

Let us look at the equality *7x + 5 = 23.* Such equality is called an *equation.* In an equation, letters stand for numbers which are necessary to find. These letters are called the *unknown or variable* in an equation. In an equation or a set of equations (simultaneous equations, for example), there can be more than one unknown or variable; however, in this present course of sixth grade mathematics, only equations with one unknown or variable will be treated.

To designate an unknown, any letter may be used. They use letters of the Latin alphabet (*a, k, n, m, p, s, t, u, x, y, z*) to designate unknowns, though the letters x and y are more frequently used. An equation is converted into a true equality by assigning a definite value of the unknown. So as we have seen above, making x = 4, our equation was converted into a true equality. Therefore, an equation is said to be an equality containing an unknown. The value of the unknown with which an equation may be converted to a true equality is called the *solution* or *root of the equation*. Some examples of equations are 5x = 10 and 3x =18. The roots of the first and second equations are 2 and 6, respectively.

2.8.2 Inequalities

Example of an inequality may be a case where the mass of 5 jars of a certain liquid is heavier or more than 15 kg. This inequality can be written in the form 5x > 15, from which it follows that x > 3. In this way, the mass of a jar with the liquid is more than 3 kg. To exactly determine the mass of one jar is considerably more difficult, for x may take on any value more than 3.

Expressions such as 5x > 15 and 5x < 15 are called **inequalities.** The variable of the variable with which an inequality is true is called the *solution* or *root of the inequality*. As we can see, the root of the inequality 5x < 15 is x < 3. Therefore, the inequality is true when the natural-number values of x are x = 1, x = 2, including x = 0. Conversely, inequality 5x > 15 has the root of x > 3, meaning that it is true when all values of x are more than 3.

If the sign > (more than) or the sign < (less than) stands between two expressions, say 8x − 5 and x + 7, then the inequality *(8x − 5 > x + 7)* or *(8x − 5 < x + 7)* is said to be *strictly defined*. And if the sign ≥ (more than or equal to) or the sign ≤ (less than or equal to) stands between two expressions, say 8x − 5 and x + 7, then the inequality *(8x − 5 ≥ x + 7)* or *(8x − 5 ≤ x + 7)* is said to be *not strictly defined*.

Examples of inequalities that are *strictly defined*:

    (a)   8x − 5 > 4x + 7
            4x > 12
            x > 3
            x = 4; 5; 6; …

    (b)   8x − 5 < 4x + 7

$4x < 12$

$x < 3$

$x = 0; 1; 2.$

Examples of inequalities that are *not strictly defined*:

(a) $8x - 5 \geq 4x + 7$

$4x \geq 12$

$x \geq 3$

$x = 3; 4; 5; \ldots$

(b) $8x - 5 \leq 4x + 7$

$4x \leq 12$

$x \leq 3$

$x = 0; 1; 2; 3.$

2.9   Scale

The relationship of the length of a line in a drawing (on a plane, figure, map, or globe) to its real or natural length is called *scale*. In other words, a scale indicates how many times the real or natural size of a represented object has been decreased. In order to represent a considerable distance on a drawing on paper, it is drawn in a decreased form to a scale by which conditionally 1 inch (or 1 centimeter, etc.) in the drawing of the figure corresponds to some given feet, yards, miles (or meter, kilometers, etc.). A scale may be indicated like this: 1 inch to 1 yard (1:36); 1 inch to 100 yards (1:3,600); 1 inch to 50 miles (1:3,168,000); and so on.

For example, if two points A and B (designating two towns) are represented on a map, we can determine the distance between them, if a scale is indicated on that map. Let us assume, for the sake of this example, that the scale on this particular map is 1 inch to 50 miles. (In reality, the scale that would be indicated on this map could be 1:3,168,000; that is, every inch on the map would correspond to 3,168,000 inches.) In order to find out the real distance between Town

A and Town B, we have to measure the distance between them on the map. Suppose that it was found by our measurement the distance between them on the map is 6 inches. To find the real or natural distance, we must multiply 6 x 50 miles = 300 miles (the actual distance between the towns).

Below we consider two analogical problems:

1. The height of Kilimanjaro, the highest mountain in Africa, is 19,340 feet. Represent the height of Mount Kilimanjaro on a drawing in your copybook, considering that 1 inch on the drawing in your copybook corresponds to 10,000 ft.

2. The length of the Nile, the longest river in the world, is 6741 kilometers. Represent the length of River Nile on a drawing in your copybook, considering that 1 cm on the drawing in your copybook corresponds to 1000 km.

2.10   Reading Problems

Knowing how to deal with or solve reading problems is as interesting as important. In your study of mathematics, you will continue from time to time to encounter reading problems of varying degrees of difficulty. Each problem contains a condition or a set of conditions in which *known* and *unknown quantities* are given, the correlation between them is described, and likewise a *question(s)* to which it would be required to give an *answer(s)*. The method of searching for the answer(s) to this question(s) is called the *solution* to the problem.

For the solution to a reading problem, you ought to study its condition(s) and to determine the hidden connection between the known and unknown quantities. For this reason, it is necessary to carry out definite reflection or *brain storming* which may lead to or be realized in two directions: from the conditions of the problem to its question, or vice versa.

As examples of reading problems, we will look at the following:

Problem 1. *To travel a certain distance, a bicyclist rode for 5 hours with the average speed of 8 mph. It still remained for him 10 miles less than the distance he had already traveled in order to reach his destination. What was the total distance that the bicyclist might have to travel?*

For reading problems, it is usually convenient to draw a diagram to represent and to simplify the conditions of the problem. Hence, the diagram in Fig. 2.12 is represented for the purpose of simplifying the solution to the problem above.

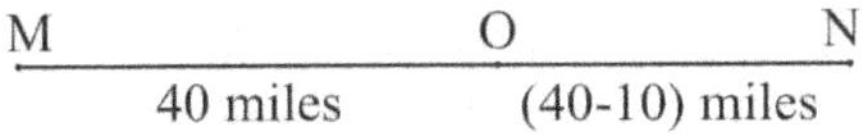

M   O   N

40 miles   (40-10) miles

Fig. 2.12. A diagram simplifying the solution to Problem 1: MO = 40 miles; ON = (40-10) miles.

First of all, we must analyze the condition of the problem. From the first sentence of the problem we can see that the bicyclist rode for 5 hours with the speed of 8 mph. By these data it is possible to determine the distance which the bicyclist had already traveled. At this point, for a pupil of average ability, his or her memory and mathematical knowledge should be activated to recognize at least that this is a *motion problem. To find the distance (d) when the speed (v) and the time (t) are given, one should simply multiply the speed by the time:* $d = v \cdot t = 8 \cdot 5 = 40$ *(miles).*

From the second sentence of the problem, it is stated and known that it still remained for the bicyclist 10 miles less than the distance he had initially covered. Therefore, we should decrease the distance he had already covered by 10 miles: *40 − 10 = 30 (miles).* In other words, knowing how many miles the bicyclist had traveled for 5 hours and how many miles yet remained for him to travel in order to complete his journey, we can determine the *total* distance of his journey: $d_t =$ *MO + ON = 40 + 30 = 70 (miles).*

It is possible to reason otherwise, starting from the question of the problem. It is required to find the total distance traveled by the bicyclist. To answer this question at once is impossible, since it is unknown how many miles the bicyclist had already traveled and how many more distance he had still to travel. Let's see if we can draw a diagram to simplify our solution, hence Fig. 2.13.

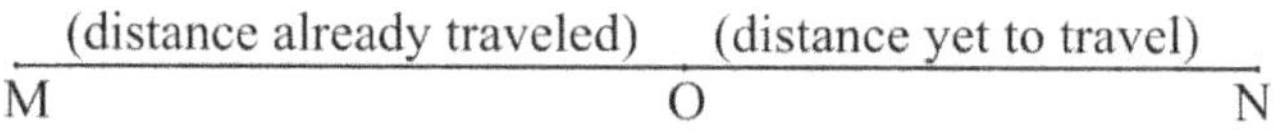

(distance already traveled)	(distance yet to travel)
M   O   N

Fig.2.13. Another diagram to simply Problem 1. MO – distance already travelled; ON – distance yet to be travelled.

In order to find the distance traveled, the speed and the time should be known. These are given in the problem. So we multiply the speed by the time to find the distance traveled: $d = v \cdot t = 8 \cdot 5 = 40$ *(miles).* The distance which the bicyclist still must travel likewise can be determined. In or to do this, we should decrease the trance traveled by 10 miles: *40 − 10 = 30 (miles).*

Consequently, the solution plan of the problem is reduced to the following steps:

    (1)   How many miles the bicyclist had traveled for 5 hours with the average speed of 8 mph?

    (2)   How many miles remained for the bicyclist to travel?

    (3)   What is the total distance the bicyclist must travel to reach his destination?

Problem 2. *A father is 45 years old and his daughter is 9 years old. By these data we can find answers to the following questions: how many years the father is older than his daughter? how many years is the daughter younger than her father? how many times the father is* *older than his daughter? how many times is the daughter younger than her father?*

To better understand the dependence between quantities given in the condition of a problem, it is frequently necessary to suggest new problems. In these new problems some unknown numerical data can be considered as known. Thus it becomes necessary to determine those which were known before. Such problems are said to be the *reverse* to the given data. Let's look at some example in problem 3.

Problem 3. *Monah typed 450 words; and Nyonoh typed 750 words in a time interval which is 10 minutes more than the time Monah spent to type 450 words. How much time did each spend typng?*

Reverse problem . Monah spent 15 minutes to type a certain number of words; and Nyonoh spent 25 minutes to also type a certain number of words. How many words did each type, if Nyonoh typed 300 more words than Monah? We compare the solutions to both problems.

Solution to Problem 3:

> **(1) 750 − 450 = 300 (words);**
> **(2) 300 ÷ 10 = 30 (words per minute);**
> **(3) 450 ÷ 30 = 15 (minutes) – the time spent by Monah;**
> **(4) 750 ÷ 30 = 25 (minutes) - the time spent by Nyonoh.**

Note: Another method of solving problem 3 is to equate the ratio of their respective times spent typing to the ratio of their numbers of words typed. If we let the time in minutes spent by Monah to be x, then we can represent the time in minutes spent by Nyonoh as $x + 10$. In this way, we can set up the solution in the following equation:

$$\frac{x}{(x+10)} = \frac{450}{750};$$

$$750x = 450(x+10)$$
$$5x = 3(x+10);$$
$$2x = 30;$$

$$x = 15.$$

Therefore, Monah spent 15 minutes to type 450 words, while Nyonoh spent (x+10 =) 25 minutes to type 750 words.

Solution to reverse problem:

(1) 25 − 15 = 10 (minutes);
(2) 300 ÷ 10 = 30 (words per minute);
(3) 30 · 15 = 450 – words typed by Monah;
(4) 30 · 25 = 750 – words typed by Nyohoh.

What other method can be used to solve the reverse problem? Is this other method easier or more difficult? Regardless of the method we use, we can see that the same numerical data are applied in the solution process of a problem and its reverse problem. It is essential to represent the conditions of these problems as the following:

Problem 3:
Monah: 450 words typed;
Nyonoh: 750 words typed, by 10 minutes longer time.
How many minutes spent by each?

Reverse problem:
Monah: 15 minutes;
Nyonoh: 25 minutes, by 300 words more.
How many words typed by each?

Problem 4. *There are 120 pupils in the 7th grade in a certain school. The number of 6th grade pupils in that school is 90 more than those in the 7th grade. How many pupils are there in the 6th grade?*

Solve the above problem and think about a reverse problem. In solving reading problems, try to make up reverse problems, and check the correctness of the answers found by the solution to their reverse problems.

Chapter Two: Questions to Test Your Understanding

1. Differentiate the following properties:
 (a) Commutative property of addition;
(b) Associative property of addition;
(c) Associative property of subtraction.

2. Answer the following:
 (a) Addition of which number to any other number gives the result of
     that very number? How may this *additive property* be formulated?
(b) Write out with the help of letters cases of addition and subtraction
    involving zero.

    3. Give examples using letters to answer the following questions:
    (a) How can we deduct a sum from a number?
    (b) How can we deduct a number from a sum?
    (c) How can we subtract a difference from a number?
    (d) How can we add a difference to a number?
    (e) How can we add a number to a difference?
    (f) How can we subtract a number from a difference?

    4. Explain how an *addition table* may be constructed.

    5. Give a few examples of the basic methods of oral calculation with
       addition and subtraction.

    6. How the *rounding off* of numbers is used in oral calculations?

    7. How the correctness of a subtraction operation may be checked?

    8. What is an *angle*, and how may it be designated?

    9. Define the following terms: (a) an open angle; (b) a right angle; (c)
       perpendicular lines; (d) compass; (e) vertex; (f) diagonal; (g) perimeter.

    10. Explain the construction of *a compass* and how it may be used to
        measured angles.

    11. Define and give example of the following: (a) obtuse angle; (b) acute
        angle; (c) complementary angles; (d) supplementary angles.

    12. What is a *polygon*? What kinds of polygons do you know?

13. What is a *triangle*, and which types of triangles can you name based on the number of their angles?

14. Differentiate the following types of triangles: (a) scalene triangle; (b) isosceles triangle; (c) equilateral triangle.

15. Define the following: (a) a rectangle; (b) a square; (c) quadrangle

16. What does it mean when two quantities are *directly proportional*?

17. What does it mean when two quantities are *inversely proportional*?

18. What two *systems of measurement* do you know?

19. Define the following terms and give an example of each: (a) expression; (b) formula; (c) equation; (d) inequality.

20. What is a *scale* and where it may be applied?

21. Give an example of a *reading problem* and its *reverse problem.*

Chapter Two: Problems and Exercises

1. Kollie was asked whether or not it was possible that the sum of $x + y$ could be equal to x. Which correct answer do you think Kollie could give to this question?

2. Calculate the following by any convenient method:
   (a) $525 + 448 + 335 + 630 + 480 + 375 + 700 + 452 + 565 + 270 + 220$;
   (b) $643 + 158 + 519 + 407 + 342 + 231$.

3. Supply the appropriate sign (=, <, >) in the place of the dots …between each given pair of expressions:
   (a) $(40 + 60) + 30$ … $100 - 30$
   (b) $(150 - 90) + 95$ … $60 + 100$
   (c) $(175 - 50) + 40$ … $200 + 40$
   (d) $(30 + 40) + 75$ … $100 + 75$

   4. Which is greater: $x - y$ or $145,000$ if $x = 350,000$ and $y = 175,000$?

5. Find the length of a broken line MNOPST, if MN = 3 inches, OP = 5 inches, and ST = 7 inches.

6. Put together an expression to solve each of the following problems and find its value:
  (a) Nelly had a certain amount of money. After buying a writing pad for $1.25, an eraser for $0.20, and a five-subject copybook for $3.50, the amount of $1.75 still remained in her possession. How much money did she have from the beginning?
  (b) Nagbe's father bought him a new bicycle. For the first week he rode 1,578 km. This distance is 605 km more than the second week of riding his new bicycle. What is the distance in kilometers did he traveled with the bicycle for the two weeks?

7. How does the sum of 527 + 381 change, if the first addend is decreased by 90 and the second is increased by 90?

8. For the first week a man traveled x miles on his new motorbike; for the second week he traveled y miles; and for the third week he traveled 350 miles more than the two previous weeks put together. How many miles did he traveled for the third week? Put together an expression for solving this problem and find its value, if $x = 975$ and $y = 705$.

9. Solve the given equations:
  (a) x + 28 = 65;
  (b) 150 + y = 208.

10. How does the difference of 567 − 312 change, if::
  (a) the minuend is decreased by 150 and the subtrahend by 50?
  (b) the minuend is increased by 150 and the subtrahend by 50?

11. Solve the following equations and check the answers.
  (a) 715 − x = 211;
  (b) y − 25 = 15;
  (c) 3x + 5 = 23;
  (d) 2y + 35 = 71.

12. By how much greater is the root of the equation 52 − (38 − x) = 75 than the root of the equation 25 + (y − 65) = 15?

13. Three brothers decided to go to the movie, but they did not have enough money. The balance of their money that remained after buying two tickets was 45 cents. They discovered that they were short of 25 cents in order to buy three tickets. How much was the cost of one ticket?

14. Sayon had $85 more than Kotee when they entered a supermarket. There in the supermarket Sayon bought a graduation present for his sister. After leaving the supermarket, Sayon had $140 less than Kotee. What was the price of the present?

15. Compare the results of the operations indicated in the following expressions, and tell which of them is greater: $534 - (312 - 199)$ or $845 - (200 + 245)$.

16. Kwesi was asked to divide 517 cents into two parts so that the number of dimes of one part is equal to the number of cents of the other part. What is the amount of money of each part?

17. A parachutist jumped from an airplane, and for the first five seconds his parachute did not open up. For the first second he descended 12 feet; and for each succeeding second he further descended 27 feet more than the previous second. Determine the distance through which the parachutist descended for the five seconds before his parachute opened up.

18. There are five packs of cigars in which there is said to be pure natural tobacco. In one of the packs, however, the cigars are not of pure natural tobacco. It is required on the first weighing to determine in which of the packs the counterfeit cigars are. The mass of a pure natural cigar is 15 gram and that of a counterfeit one is 10 grams. In order to find the answer, one cigar is taken from the first pack, two cigars from the second pack, three from the third, four from the fourth, and five from the fifth. The total mass of all the cigars taken is 200 grams. *In which of the five packs are the counterfeit cigars? Explain why this method of weighing enables us to find the answer to this question.*

19. A school bag, a copybook, and a textbook were bought for the amount of $135. The school bag and the copybook together cost $81; and the copybook plus the textbook together cost $59. What is the separate cost of each item?

20. Answer the following questions:
  (a) Find the greatest number which is the sum of three greatest four-digit numbers.

110

(b) Find the smallest number which is the difference of the two smallest three-digit numbers.

(c) Find the number which is the difference between the greatest two-digit number and the smallest two-digit number.

21. Calculate the following by a method convenient to you:
(a) 8 + 14 + 20 + 26 + 32 + 38 + 44 + 50;
(b) 10 + 20 + 30 + ... + 80 + 90 +100;
(c) 4 + 8 + 12 + ... + 392 + 396 + 400.

22. Instead of the dots, write the name of each relevant unit that makes each equality to be correct.
(a) 10 yards + 7.5 feet + 9 ... = 12 yards + 2 feet + 3 inches;
(b) 5 tons + 500 kg − 800 ... = 4 tons 700 kg;
(c) 15 km + 550 ... = 15 km 55 meters;
(d) 18 dollars + 25 cents + 85 ... = 19 dollars + 15 cents.

23. Calculate the following orally:
(a) (i) 165 + 130; (ii) 165 + 45; (iii) 165 + 145;
(b) (i) 1,000 − 850; (ii) 1,250 − 850; (iii) 1,500 − 850.

24. Find the unknown x and tell how it is called in each example:
(a) 161 + x = 196;
(b) x + 405 = 801;
(c) 250 +319 = x;
(d) 159 − x = 45;
(e) x − 85 = 95;
(f) 680 − 205 = x.

25. Ali spent $290 within four days. On the first day he spent $39.25; on the second day he spent $51.75; on the third day he spent $23. (a) How much money remained with him after the second day? (b) How much money did he spend on the fourth day?

26. Explain the method used in each of the following calculations:
(a) 567 − 317 = 570 − 320 = 250;
(b) 435 − 185 = 450 − 200 = 250;
(c) 1,248 − 395 − 421 = 1248 − (395 + 421) = 432;
(d) 215 + 485 = 215 + 500 − 15 = 700.

27. From the sum of the numbers 9,538, 761 and 3,298,775 deduct their difference. By which other method can we solve this problem?

28. From the sum of 65,412 and 32,145 deduct the difference of the two numbers consecutively following after them, respectively.

29. In each of the following addition and subtraction operations, supply the missing digit represented by each asterisk.

(a)
```
    7 * 5 * 3 *
  + * 5 * 9 * 5
    9 5 4 2 1 6
```

(b)
```
    6 * 5 * 4 * 3
  - 1 2 * 5 * 9 *
    * 5 6 4 3 3 4
```

(c)
```
      3 5 7 6
      5 6 4 2
  + * * * *
    1 3 7 8 5
```

(d)
```
      * * * *
      9 5 7 3
  +   4 0 9 8
    1 4 9 0 5
```

30. A farm was planted with rice, eddoes, and yams. 203 hectares of farm land was planted with rice; eddoes – by 95 hectares less than rice; and yams – by 38 hectares less than eddoes. Find the total area of farm land that was planted.

31. Calculate the following by the most convenient method:
(a) 527 + 305 + 273;
(b) 1581 + 4549 + 519;
(c) 1987 + 2853 + 313 + 547;
(d) 843 + 950 + 557.

32. With the help of a ruler, draw an acute and obtuse angles.

33. With the help of a protractor, measure the angles represented in Fig.2.14 below:

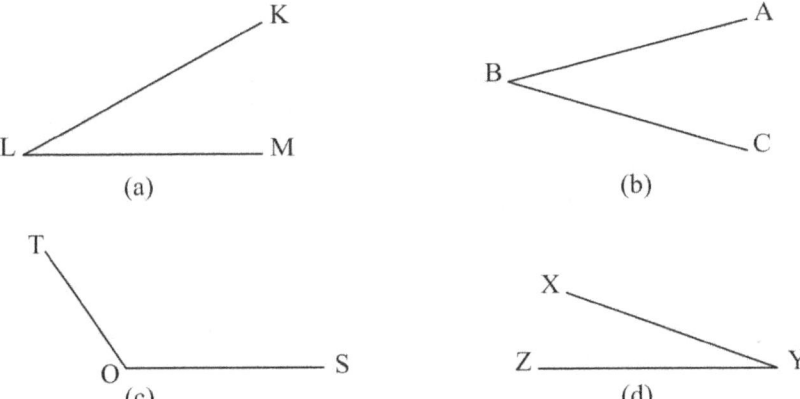

Fig. 2.14. Measurement of angles using a protractor.

34. Draw two angles and designate their vertices and sides.

35. Draw the following figures:
(a) acute angle ∠ABC
(b) right angle ∠XYZ
(c) obtuse angle ∠DEF
(d) open angle ∠MON

36. Which angle is formed by the minute and hour hands on a watch at the following times?
(a) 15 hours;
(b) 6 hours;
(c) 10 hours;
(d) 9 hours.

37. What fractional part of an open (straight) angle is angle D, if:
(a) $∠D = 90^0$;
(b) $∠D = 30^0$;
(c) $∠D = 45^0$;
(d) $∠D = 60^0$.

38. A watch shows 18 hours. What time will it show if the minute hand shifts by:
(a) 360^0;
(b) 180^0;
(c) 90^0;

113

(d) 60^0?

39. Draw two equal obtuse angles having a common side, and the two other sides to form an angle of 60 degrees. How many degrees does each obtuse angle contain?

40. Name at least three properties of a rectangle. Draw a rectangle with length of 2 inches and width of 1 inch. Find the perimeter of this rectangle.

41. By the graphical representation in Fig.2.15 below, calculate the perimeter of each of the following:
(a) Equilateral triangle ΔABE;
(b) Quadrilateral ABCE, and all other identifiable quadrilaterals;
(c) Isosceles triangle ΔCED.

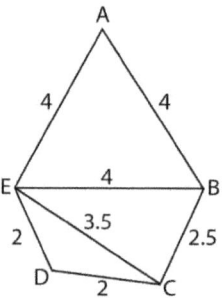

Fig.2.15. Calculation of perimeters of polygons.

42. With the help of a ruler, a pair of compasses, and a protractor, draw a triangle DCE two sides of which are 2 inches each and the angle between them is 45^0. Measure the other two angles and the length of the third side of this triangle.

43. Find the base of an isosceles triangle perimeter of which is equal to 7 inches and each of the lateral sides is equal to 2 inches. Draw this triangle.

44. Perimeter of which polygon is greater: the triangle or the rectangle in Fig.2.16? By how much greater?

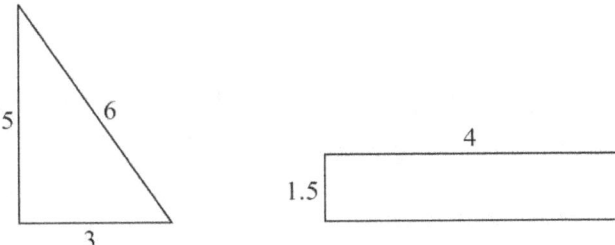

Fig.2.16. Determining and comparing the perimeters of polygons.

45. Draw a right triangle, an acute triangle, and an obtuse triangle. Using a protractor, find the sum of degree measurements of the angles in each triangle.

46. Draw an arbitrary hexagon. Measure the length of each side and find the perimeter of this polygon.

47. The sum of the lengths of the first and second sides of a triangle is 7 cm; the sum of the lengths of the first and third sides is 8 cm; and the sum of the lengths of the second and third sides is 9 cm. What do you think about the sum 7 cm + 8cm + 9 cm? Find the perimeter of this triangle and the length of each side.

48. Write the formula for finding the perimeter of a triangle with sides x, y, and z (in inches).

49. One side of a triangle is 15 in, and the second side is 10 in, as shown in Fig.2.17 below. What may be the value of the third side represented by the letter y?

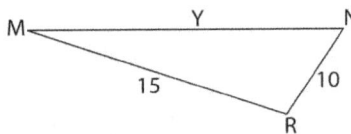

Fig.2.17. Finding the length of the third side of a triangle when the lengths of two sides are known.

50. Supply the missing values in Table 2.6 below.

115

Distance, miles				150	150	150	150
Speed, mph	35	45	90	70	60	50	40
Time, hours	5	5	5				

Table 2.6. Calculations involving *distance, speed,* and *time.*

(a) What is the nature of dependence between speed and time?
(b) What is the nature of dependence between speed and distance?
(c) What is the nature of dependence between distance and time?

51. Two cars simultaneously left from a car parking station and went in opposite directions. Which of them will be the first to arrive at its destination, if it is required for the first one to travel a distance three times more than the second, and goes three times faster.

52. Express the following in centners:
(a) 5 tons 200 kg;
(b) 3 tons 3 centners;
(c) 900 kg.

53. Perform the indicated operations:

(a)　　41km　21m
　　+　18km　81m

(b)　　15kg 50g
　　−　10kg 80g

(c)　　5 hours 180 sec
　　+ 2 hours 240 sec

54. Supply the missing values in *miles, mph,* and *hours* in Table 2.7 below.

Distance, miles		1415 miles	1750 miles	840 miles		3015 miles
Speed, mph	32 mph			70 mph	45 mph	15 mph
Time	4 hours	5 hours	50 seconds		30 minutes	

Table 2.7. Calculations involving *distance, speed,* and *time*

55. Find the value of the expression $x + 17,718$ if $x = 25,025$.

56. Calculate the value of the expression $49,350 - x + 21,505$ if $x = 36,125$.

57. A man is 46 years old. He is 32 years older than his son. What is the age of his son?

58. What is the result to the expression $x + x \div 7$, if $x = 84$.

59. Find the sum and difference of the following quantities:
(a) 21km 25m and 10km 650m;
(b) 5 ton 400 kg and 2 ton 700kg;
(c) $905.68 and $509..86.

60. In one kinja (a type of country basket), there are x number of golden plums; and in the other kinja, there are 11 golden plums less than those in the first. How many golden plums in both kinjas, if $x = 38$.

61. Solve the following sets of inequalities. How many solutions does each inequality contain?
(a) (i) $x + 4 < 7$;   (ii) $x + 4 \leq 7$;   (iii) $x + 4 > 7$
(b) (i) $18 - x < 8$;   (ii) $18 - x \leq 8$;   (iii) $18 - x > 8$.
(c) (i) $3x < 51$;   (ii) $3x \leq 51$;   (iii) $3x > 51$;
(d) (i) $\frac{55}{x} < 11$;   (ii) $\frac{55}{x} \leq 11$;   (iii) $\frac{55}{x} > 11$;

Note: The roots or solutions to the given inequalities should be *natural numbers* only and the number *zero.* The following should therefore be taken into consideration:
- *the difference $x - y$ is reasonable and can be defined only when and if $x \geq y$;*
- *it is impossible or not allowable to divide by zero;*
- *an inequality (i.e. the sign of an inequality) is reversed when and if its both sides are multiplied by a negative.*

62. Solve the following equations:
(a) $x + 342 = 507$;
(b) $x - 108 = 250$;
(c) $(a - 785) - 605 = 2,567$;
(d) $(810 + y) + 452 = 1,534$

(e) $5x + 3 = 38$;
(f) $50 \div x = 5$;
(g) $y \div 4 = 100$.

63. Set up the relevant equation and find the unknown numbers.
   (a) If an unknown number is divided by 3 and the quotient is decreased by 5, then we get 16 as the answer.
   (b) The difference of 450 and an unknown number was decreased by 10 times, and the result became 25.

64. What is the value of the variable (an unknown number) in which the equality $5 + 4x = 25$ will be true?

65. Put together an expression with the help of letters for solving each of the following problems:
   (a) A market woman sells 7 oranges for 21 cents. At that rate, how many oranges can be bought for d cents? How many oranges if $d = 42$ cents?
   (b) An electronic mimeographing machine can produce 64 mimeographs in 4 seconds. At that rate, how many mimeographs can be produced in t seconds? What if t is equal to 22 seconds?

66. 1.5 yams were placed on a pan on one side of a scale. On the other side of the scale 0.25 yam and a weight of 5 kg were placed in order to balance the scale. What is the actual weight of this yam? (**NOTE:** $1.5 = 1\frac{1}{2}$; $0.25 = \frac{1}{4}$)

67. The size of a rectangular area is 60 meters by 100 meters. Draw a plan of this area with the scale: 1cm = 20m.

68. Consider Fig.2.16 below. Making use of the scale indicated, solve the given problem. An airbus was traveling from Guinea Conakry over Monrovia (Liberia) and Abidjan (Ivory Coast) to Abuja (Nigeria).
   (a) How many kilometers in all the airbus flew?
   (b) How many kilometers from Guinea Conakry to Monrovia?
   (c) How many kilometers from Monrovia to Abidjan?
   (d) How many kilometers from Abidjan to Abuja?

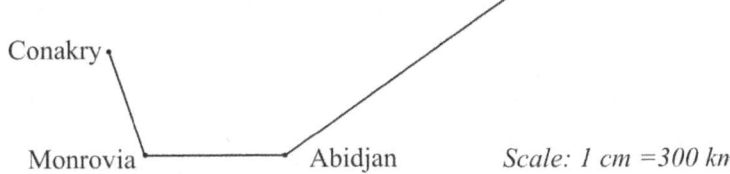

Conakry

Monrovia   Abidjan   *Scale: 1 cm =300 km*

Fig.2.18. Using a scale to determine distance on a map.

69. The length of the St. Paul River is 4,200 miles and the Sanquin River is 1,200 miles less than it. Draw in your copybook the lengths of these rivers using line segments with the *scale: 1 inch = 600 miles*.

70. Solve the following problems and make up a *reverse problem* to each.
    (a) The weight of a bag of rice is 50 kg, and the weight of a bag of farina is 42kg 580g. By how much is the bag of rice heavier than the bag of farina?
    (b) Tenesee picked 96 plums from a plum tree and gave some of them to Tanneh. How many plums did he give to her, if 65 plums still remained with him?

71. In preparation of Tundy's graduation party his father bought 175 bottles of beer. The soft drinks he bought were 20 bottles less than those of beer; and the wine was 63 bottles less than those of the soft drinks. How many bottles of wine did Tundy's father buy?

72. The total number of tourists in two buses was 79. When 11 tourists came out of the second bus, there still remained in it 17 tourists. How many tourists were in the first bus?

73. The mass of 12 bananas is equal to the mass of 2 pineapples plus 2 butter pears. The butter pear is two times lighter than the pineapple. How many bananas can we take so that their mass is equal to the mass of one pineapple?

74. In seven big boxes and five small ones there are 124 copybooks. In other three big boxes and five small ones there are 76 copybooks. How many copybooks are there in each of the big boxes?

75. Considering problem number 74 above, how many copybooks are there in each small box?

76. In the warehouse storage of a firm, there were 85,686 cartons of goods for sale on a Christmas Eve. Before lunch time 14,345 cartons were sold. After lunch additional 26,341 cartons were brought into the storage. How many cartons of goods remained in the storage?

77. 10,526 people work in a supermarket, among them 3,025 women. Calculate how many more men than women working in this supermarket.

78. Make up a *reverse problem* to problem number 77.

79. They brought 205 chicks, ducks, and geese to a poultry farm. The chicks and geese together were 99; and the chicks and ducks together were 131. How many chicks, ducks, and geese were there separately?

80. Name few values of the variable x by which the expression $130 - x > 123$ would not be true.

81. If k is equal to 30, find the value of each of the following expressions:

(a) $(45 + k) \div 25$;
(b) $k \div 5 - k \div 15$.

82. Solve the following equations:
(a) $502 - 2x = 122$;
(b) $3x - 15 = x + 47$;
(c) $60 - (x - 5) = 25$.

83. Solve the following problems, making up the relevant expressions:
(a) On an experimental plot of farm land was sown 120 kg of rice. Corn, cassava, and eddoes were also sown on the plot. The quantity of corn was 4 times less than that of rice; the quantity of cassava was 3 times less than the quantity of rice and corn together; and the quantity of eddoes was 2 times less than the quantity of corn and cassava together. Find the quantity in kilograms of each crop, and their total quantity.
(b) For 5 days a farmer planted 200 kg of seed rice on her farm. How many kilograms of seed rice could she plant for 4 days, if it is known that for every following day she would plant x kg more?

84. Doe and his girlfriend Juagbeh started walking at the same time towards each other for their appointment at lunch time. Doe started from the Telecommunications building on Lynch Street; and Juagbeh started from the Finance Ministry building at the intersection of Broad & Mechlin Streets. After 5 minutes, Juagbeh was more far from the Telecommunications building than was Doe from the Finance Ministry building. Which of the two was walking faster?

85. The tens digit (that is, the digit to the left) of a two-digit number is 5. The digit 3 is placed between the digits of this number. By how much is the resulting three-digit number greater than the initial two-digit number?

Chapter Two: Exercises to Rate your Ability

1. There was 105,648 kg of goods in five warehouses of a company. 3,502 kg of goods were added to the fifth warehouse; 21,009 kg of goods were taken away from the fourth; 52,319 kg of goods were taken away from the third and added to those of the second warehouse. Additional 63,989 kg of goods were taken away from the first warehouse. What quantity of goods remained in all of the five warehouses put together?

2. Find the sum of the roots of the equations: $(a - 3) + 5a = 69$ and $(6k + 20) - 3k = 80$.

3. Solve the equation: $(x + 300) + 5x = 715 + x$.

4. Find the value of the expression: $(k - 480) + 153$, if $k = 589$.

5. Solve the following equations:
    (a) $(401 - k) + 159 = 509$;
    (b) $k - 100 = 31$;
    (c) $193 - k = 51$.

7. Execute the indicated operations:
    (a) $2,013 - 1,539$;
    (b) $81,312 - 9,091$;
    (c) $50 + 200 \div 5$.

8. Make up the equations and find the unknown in each:

(a) If an unknown number is divided by 3 and the quotient is decreased by 100, then the   result is 5.

(b) The difference between 80 and an unknown number was decreased by 2 times and the result was 30.

CHAPTER THREE: MULTIPLICATION AND DIVISION

3.1   An Overview of the Concepts

It is important to be reminded that multiplication of natural numbers is a particular kind of addition, when it is necessary to find the sum of the same addends. For example, in order to find out how many candles are there in 6 boxes, each of which containing 8 candles, it is necessary to add: $8 + 8 + 8 + 8 + 8 + 8 = 48$ *(candles)*. It is quite convenient particularly if there are many addends. So it is generally accepted to write in a shorter way like this: $8 \cdot 6 = 48$ (candles). In the shorter notation, 8 is the repeated addend, and 6 is the number of times it has been repeated. In a multiplication operation they are correspondingly referred to as multiplicand and multiplier, and the result is known as the product. The multiplicand and multiplier are frequently called factors. The product of two factors which are the same, for instance $7 \cdot 7$, is written like this: 7^2 (seven squared). Obviously, $7^2 = 7 \cdot 7 = 49$. The product of three factors which are the same, for example, $11 \cdot 11 \cdot 11$, is written like this: 11^3 (eleven cubed, or eleven raised to the third power). It is also obvious that that $11^3 = 11 \cdot 11 \cdot 11 = 1331$. These concepts were introduced in the beginning section of this book.

Let us assume we have 48 candles placed equally into 6 boxes. It is not hard to recognize that 8 would be in one box. This means we have shared all candles into 6 equal parts; i.e. we have performed an operation of division: $48 \div 6 = 8$ (candles). There is another case when it is known that 48 candles have been placed into some boxes by 8 candles in each. It is understandable that the quantity of boxes would be $48 \div 8 = 6$ (boxes), i.e. again we have performed a division operation.

In each case, it was necessary to determine or find the other factor by the known product and one of the factors. Therefore, division is an operation by means of which it is possible to determine the second factor by a known product and one of the factors. In the process of division, the first number (that is, the number which is being divided) is called *dividend*. The second number (that is, the number by which the dividend is divided) is called *divisor*. And the result of a division operation is called *quotient*.

Division is an operation which is reverse to multiplication, in the same way subtraction is opposite to addition. Therefore, the following statements are true:

1. The finding of an unknown addend by a known sum and another addend is said to be an operation reverse to addition (i.e. subtraction). For example, $15 - 5 = 10$. In each case, a known sum and a known addend are used to determine the unknown addend. Here, 15 is the known sum, 5 is a known addend, and 10 is the unknown addend which has been determined.
2. The finding of an unknown factor by a known product and another factor is said to be an operation reverse to multiplication (i.e. division). For example, $24 \div 6 = 4$. In each case, the known product and a known factor are used to determine an unknown factor. Here, 24 is the known product, 6 is the known factor (or divisor), and 4 is the unknown factor (or quotient).

Based on the foregoing concepts, the following rules are relevantly applicable:

1. In order to find an unknown factor, it is necessary to divide the product by the known factor. In other words, in order to solve an equation of the type $3x = 60$, it is required to perform the division $x = 60 \div 3 = 20$.
2. In order to find an unknown divisor, it is necessary to divide the dividend by the quotient. In other words, in order to solve the equation of the type $60/x = 3$, it is required to perform the division $x = 60/3 = 20$.
3. In order to find an unknown dividend, it is necessary to multiply the divisor by the quotient. For instance, *if* $x / 6 = 3$, *then* $x = 6 \cdot 3 = 18$. 18 is the dividend, 6 is the divisor, and 3 is the quotient.

Let us examine particular cases of multiplication. We multiply the number 1 by the number 9. For this, we find the sum $1 + 1 + 1 + 1 + 1 + 1 + 1 + 1 + 1 = 9$. That is why $1 \cdot 9 = 9$. By analogy, in order to multiply 1 by a certain number x, it is necessary to add the number 1 x times. Therefore, $1 \cdot x = x$. If we multiply the number 0 (*zero*) by x, then we will have the answer to be 0 (*zero*): $0 \cdot x = 0$.

In as much as the operation of division is reverse to that of multiplication, the following rule is true:

To divide the number m by the number n means to find an unknown number x, such that $x \cdot n = m$. For example, $27 \div 3 = 9$, and $9 \cdot 3 = 27$.

The following examples convince us that for any natural number, the equalities below are true:

$$m \div n = x, \text{ since } x \cdot n = m;$$
$$m \div m = 1, \text{ since } 1 \cdot m = m;$$
$$m \div 1 = m, \text{ since } m \cdot 1 = m;$$
$$0 \div m = 0, \text{ since } 0 \cdot m = 0;$$
$$m \div 0 \text{ (is undefined; does not exist).}$$

REMEMBER!!! IT IS NOT POSSIBLE TO DIVIDE BY ZERO; THAT IS, ZERO CANNOT BE A DIVISOR.

3.2   Multiplication and Division of Natural Numbers

The decimal system of notation or counting has the property that 10 units of any rank make up 1 unit of the neighboring higher rank. If for instance the number 5 is multiplied by 10, then we obtain the result of five units of the next higher rank, i.e. 5 tens. If the number 160 is multiplied by 10, then we obtain the result of 1,600. The number 160 comprises of 1 hundred and 6 tens, while the number 1,600 comprises of 1 thousand and 6 hundreds. It is interesting to note that by multiplying 160 (or 1 hundred and 6 tens) by 10 we obtain 1,600 (or 1 thousand and 6 hundreds). 1 thousand is the next higher place-value rank after 1 hundred; and 6 hundreds is the next higher place-value rank after 6 tens. This notational and numerical phenomenon is very unique about the decimal system.

Pupils at an earlier stage should be able to understand, appreciate, and use the unique properties of the decimal system to their advantage so as to enhance their abilities in operating with numbers.. This is another useful example. If the number 546 is multiplied by 100, then we have the result of 54,600. Each digit of the number 546 has shifted or been displaced by two ranked units to the left, with the two last digits being zeros.

In general, in order to multiply a natural number by such ranked units as 10, 100, 1000, and so on, it is necessary to write to the left of that number as much zeros as are in the ranked unit by which the number is being multiplied.

To divide by ranked units (10, 100, 1000, and so on), one should throw off or discard in the dividend as many zeros as are in the ranked unit. In reality, this means that one should move or shift the decimal point to the left as many places as are the number of zeros in the ranked units. For example, $34,000 \div 10 = 3,400$; $26,000 \div 100 = 260$; and $4,900 \div 1000 = 4.9$.

3.3   Commutative Property of Multiplication

The rectangle in Fig.3.1 is broken up or laid out into squares. It is easy to observe that it is possible to find the number of squares by three methods. The first method is to count the number of squares in one row and then multiply by the total number of rows, i.e. $12 \cdot 5 = 60$. The second method is to count the number of squares in one column and then multiply by the total number of columns, i.e. $5 \cdot 12 = 60$. The third method is to count the number of squares in a row and the number of squares in a column, and then to multiply the two numbers: $12 \cdot 5 = 60$. In each case, we have been able to determine the same number of squares; therefore, the results are the same. This is a property of multiplication which

Fig. 3.1. A rectangle

can be generalized by the following expression: $x \cdot y = y \cdot x$. This property is true with any values of factors. Irrespective of the positions of the factors the product does not change. In other words, the result of a multiplication operation does not change from the re-arrangement of its factors. This is called the commutative property of multiplication. This property is also applicable in addition. We remember from section 2.1 of this book that in addition it is possible to transpose or re-arrange the addends and still arrive at the same sum. This is called the commutative law or property of addition, which may be formulated as: $x + y = y + x$. The commutative properties of addition and multiplication are very important and useful tools in addition and multiplication operations. They may jointly be formulated like this: "the product (or sum) of a multiplication (or addition) operation does not change from the re-arrangement of its factors (addends)."

Since $x \cdot 1 = x$, then it is possible to formulate such a rule that if one of two factors is equal to 1, then the product is equal to the second factor. There is a particular case when both factors are equal to 1. For example, $1 \cdot 1 = 1$. If one (or both) of the factors are equal to zero, then the product is also equal to zero. For example, $x \cdot 0 = 0$; $0 \cdot x = 0$; and $0 \cdot 0 = 0$.

3.4   Associative Property of Multiplication

Problem: A case contains 5 rows of 7 jars of pineapple juice in each row. The capacity of each jar of pineapple juice is 3 kg. What is the total capacity of all the jars?

Solution: We first of all determine the capacity of 7 jars in one row: $7 \cdot 3$ (kg); then the capacity of all the jars in 5 rows will be: $(7 \cdot 3) \cdot 5 = 105$ (kg). It is possible to determine the capacity of all the jars otherwise: first determine the capacity of 5 jars: $5 \cdot 3$ (kg); then the capacity of all the jars (with 7 in each row) will be: $(5 \cdot 3) \cdot 7 = 105$ (kg).

Consequently: $(7 \cdot 3) \cdot 5 = (5 \cdot 3) \cdot 7 = 105$. This is true for any other natural numbers: $(2 \cdot 3) \cdot 5 = 2 \cdot (3 \cdot 5)$; $(6 \cdot 7) \cdot 9 = 6 \cdot (7 \cdot 9)$. In general, for any values of k, m, and n, the following expression is true:

$$\boxed{(k \cdot m) \cdot n = k \cdot (m \cdot n) = (k \cdot n) \cdot m}$$

The above expression is called the *associative property of multiplication,* which states that *"in order to multiply the product of two numbers by a third number, it is necessary to multiply the first number by the product of the second and third numbers."* The associative and commutative properties of multiplication allow us to join or group together three factors during multiplication in different ways without changing the product: $7 \cdot 2 \cdot 5 = 7 \cdot 5 \cdot 2 = 7 \cdot 10 = 70$. Associative and commutative properties of multiplication are extended to a larger or greater number of factors. For example, $12 \cdot 35 \cdot 5 \cdot 2 = (12 \cdot 5) \cdot (35 \cdot 2) = 60 \cdot 70 = 4,200$.

3.5   Distributive Property of Multiplication and Division

Problem: A case contains 4 rows of bottles of soft drinks, with 13 bottles in each row. Eight of the 13 bottles in each row are Coca-cola, and the remaining five are Fanta bottles. What is the total number of bottles in the case?

Solution: In order to find the total number of bottles in the case, it is necessary to determine the number of bottles in one row and multiply it by the number of rows. In this way, the result will be: *(8 + 5) · 4 = 52*. It is possible also to determine how many Coca-cola bottles (8 · 4 =32) and how many Fanta bottles (5 · 4 = 20), and then add them together. Therefore, the total number of bottles in the case will be *32 + 20 = 52*. The results are the same; in both cases we have determined the same number of soft drink bottles: *(8 + 5) · 4 = 8 · 4 + 5 · 4 = 52*. It is possible to check to be convinced, using other numbers, for example:

1. $(7 + 8) \cdot 5 = 7 \cdot 5 + 8 \cdot 5 = 75$;
2. $(6 + 7) \cdot 12 = 6 \cdot 12 + 7 \cdot 12 = 156$;
3. $(15 - 8) \cdot 9 = 135 - 72 = 63$;
4. $(15 - 8) \cdot 9 = 7 \cdot 9 = 63$.

In general, for any numbers k, m, and n the following expression is true:

$$(k + m) \cdot n = k \cdot n + m \cdot n;$$
$$(k - m) \cdot n = k \cdot n - m \cdot n$$

The above expression is called the ***distributive property of multiplication***, which states that *"in order to multiply a sum (or difference) by a given number, it is necessary to multiply each addend (or the minuend and subtrahend) by that number and then to add (or subtract) the resulting products."* This property may also be extended to division. It is possible to formulate an analogical law that *"in order to divide a sum (or difference) by a given number, it is necessary to divide each addend (or the minuend and subtrahend) by that number and then add (or subtract) the resulting quotients."* This property is called the *distributive property of division*; and it may be summarized as given below:

$$(k + m) \div n = k \div n + m \div n;$$
$$(k - m) \div n = k \div n - m \div n$$

These properties or rules are used to open parentheses, simplify expressions, and check the correctness of performing arithmetic operations.

Some examples:

(a) $65 \cdot 3 = (60 + 5) \cdot 3 = 180 + 15 = 195;$
(b) $98 \cdot 5 = (100 - 2) \cdot 5 = 500 - 10 = 490;$
(c) $248 \div 8 = (152 + 96) \div 8 = 19 + 12 = 31;$
(d) $105 \div 3 (60 + 45) \div 3 = 20 + 15 = 35;$
(e) $639 \cdot 4 = (600 + 30 + 9) \cdot 4 = 2400 + 120 + 36 = 2{,}556;$
(f) $7x + 6x - 5x = (7 + 6 - 5) x = 8x.$

3.6   Written Multiplication Operation with Natural Numbers

All cases of multiplication of simple one-digit numbers by simple numbers are presented in multiplication tables which are studied by pupils in beginning forms or classes. It is possible and simpler to multiply smaller numbers orally. For the multiplication of larger numbers, they make use of the rules of multiplication in column.

Let us suppose it is required to multiply 1296 by 435. We can break down the number 435 into its place-value units (called *ranked units*) and make use of the distributive property of multiplication. We will obtain the following result: $1296 \cdot 435 = 1296 \cdot (400 + 30 + 5) = 1296 \cdot 400 + 1296 \cdot 30 + 1296 \cdot 5 = 518{,}400 + 38{,}880 + 6{,}480 = 563{,}760.$

The above operation of multiplication, involving the breaking down of 435 into addends (according to the place values of its component digits) and the addition of calculated products, is reproduced below in a more simplified form:

```
         1296
       x  435
         6480  ←————————  1296 x 5 =    6480
         3888  ←————————  1296 x 30 =  38880
     +   5184  ←————————  1296 x 400 = 518400
        563760
```

It should be observed that the zeros in the second and third lines are implied, but they are not written. It makes no difference if they are written; the result remains the same.

The conclusion that can be drawn from the preceding multiplication operation is that natural numbers are multiplied on the basis of their place values, starting from the ranks of units, tens, hundreds, and so on, and then adding the resulting products.

Let us examine a case where a zero is found in one or some ranks of the multiplier. As it is already well known, in multiplying by zero, the result is always zero. Therefore, in multiplying in a column, one or more lines (or rows) would be made up of only zeros, where a zero or zeros are present in one of the factors. As we have said above, these zeros have no influence on the result of the addition; and it is not only possible, but necessary to omit the line (s) or row (s) made up of only zeros, as in the example below.

```
        54067                    54067
      x  406                   x  406
       324402                   324402
        00000                 + 216268
    + 216268                   21951202
     21951202
```

If one or two of the factors end in zeros, then they don't multiply these zeros; they simply are put aside and later added to the result of the multiplication of the non-zero digits, as in the examples below:

(a)   35024
        x 2300
      105072
    + 70048
      80555200

(b)   58900
        x 320
      1178
    + 1767
      18848000

*Please note that 35024 x 2300 is equivalent to 35024 x 23 and just adding the two zeros at the end of 2300 to the product of **35024** and **23**. In the same way, 58900 x 320 is equivalent to 589 x 32 and just adding three zeros (two at the end of 58900 and one at the end of 320) to the product of **589** and **32**.*

3.7   Written Division Operation with Natural Numbers

From previous classes, we remember how division in column is performed. Division of any natural number by a natural number can be performed; and it follows that the dividend should be greater than the divisor.

Let us look at the following division operation of numbers, and explain how each digit in the quotient was obtained. Also we should pay attention to the zeros in the quotient of the second example, and understand why they do stand there.

In order not to make an error in the division of many-digit numbers, it is expedient to make use of a *method of approximation* in determining the number of digits in the quotient. Here, we give two examples with some explanation

(1) $934,735,842 \div 36,459 = 25,638$

$$
36459 \overline{)934735842} \quad {}^{25638}
$$

```
          25638
36459 )934735842
    -72918
     205555
    -182295
     232608
    -218754
     138544
    -109377
     291672
    -291672
          0
```

(2) $848,510,193 \div 27,906 = 30,406$

```
          30406
27906 )848510193
    -83718
     113301
   - 111624
     167793
    -167436
        357
```

In the first example, the first five digits in the dividend form the number 93473 which when divided by 36459 gives the digit 2 in the quotient; actually, it is simpler to reckon that the digit 2 is obtained from dividing 93 by 36. The four remaining ranked units (5842) in the dividend when added to the corresponding remainders (one at a time after each subtraction) give additional four digits 5638 in the quotient. Consequently, the quotient comprises of five digits, that is, a five-digit number; it is roughly equal to 20000. Such a method is convenient when the quotient has zeros in between as in the second example. In this way, we determine that five digits will be in the quotient of the second example.

In the first example, the number 934,735,842 was divided by 36,459 and we obtained the number 25,638 as the quotient. It can be said that the number 934,735,842 was divided by 36,459 *without remainder.* In the second example, in dividing the number 848,510,193 by 27,906 we have found the quotient to be 30,406 and still remained the number 357. It is not possible to continue the division operation because 357 is less than the divisor 27,906. Hence, the number 357 is called the *remainder* from the division of the number 848,510,193 by the number 27,906; and the quotient 30,406 is said to be *incomplete.* Such a division is called *division with a remainder.* If in dividing one number by another we are left with a remainder, then we say that the first number cannot be divided by the second without a remainder. Thus it can be seen that one natural number is not always divisible by another without a remainder. In the case of division with a remainder, the remainder is always less than the divisor.

Let us look at a simple problem in which it is required to equally divide 33 Christmas balloons among 10 boys. It is obvious that each boy will get 3 balloons, with 3 balloons still remaining. It follows from here that the dividend 33 is equal to $10 \cdot 3 + 3$ (that is, to the sum of the product of the divisor 10 and the incomplete quotient 3, plus the remainder 3).

In general, if a dividend is equal to m and the divisor to n, then the quotient q and the remainder r are determined such that $\mathbf{m = n \cdot q + r}$, where $r < n$. If $r = 0$, then it is said that the dividend m can be divided without remainder by the divisor n, as well as by the quotient q, since $\mathbf{m = n \cdot q}$.

Let us look at some features of division. We can figure out how the quotient changes if the dividend is increased (or decreased) x times, or if the divisor is increased (or decreased) x times. In general, it is easy to observe that:
- *the quotient increases (or decreases) the same number of times the dividend increases (or decreases);*

- *the quotient increases (or decreases)the same number of times the divisor decreases (or increases);*
- *The quotient does not change (remains the same) if both dividend and divisor are increased (or decreased) the same number of times; in other words, multiplying or dividing both dividend and divisor by the same number does not change the value of the quotient.*

These properties of division are very important and very useful in the solution of many problems.

3.8 Multiplication of Rational Numbers

From previous arithmetic classes or from topics already covered in this book you can find the product of two positive numbers. Here we deal with problems involving how to find the product of two negative numbers, the product of two numbers with different signs, and likewise those concerning the application of rules for the multiplication of rational numbers. Let us look at the following example problems below:

Problem 1.

(a) $(-9) \cdot (+2) = -18$;

(b) $(+6) \cdot (-4) = -24$;

(c) $(-7) \cdot (-3) = +21$;

(d) $(-8) \cdot (+5) = -40$;

(e) $10 \cdot (-12) = -120$;

(f) $(+5.8) \cdot (-3.2) = -18.56$;

(g) $(-7.4) \cdot (+6.1) = -45.14$.

Problem 2.

(a) $(-\frac{3}{5}) \cdot (+\frac{5}{12}) = -\frac{1}{4}$;

(b) $(+\frac{6}{7}) \cdot (-\frac{8}{9}) = -\frac{16}{21}$;

(c) $(-\frac{2}{3}) \cdot (-\frac{4}{5}) = +\frac{8}{15}$;

Problem 3.

(a) $(-5.7) \cdot 0 = 0$;

(b) $0 \cdot (+2.5) = 0$;

(c) $24.9 \cdot p = 0$;
$p = 0$;

(d) $4.3 \cdot (a + 7.8) = 0$;
$a + 7.8 = 0$;
$a = -7.8$;

(e) $(m - 3.7) \cdot 5 = 0$;
$m - 3.7 = 0$;
$m = 3.7$;

(f) $|y + 1.6| = 0$;
$y + 1.6 = 0$;
$y = -1.6$;

(g) $|d - 8.9| \cdot 10.8 = 0$;

$|d - 8.9| = 0$;

$d = |8.9|$.

From the solutions to the problems above, we can formulate 3 rules for the multiplication of rational numbers:

1. *The product of two numbers with the same signs is a positive number; and the module of said product is equal to the product of the modules of the factors.*

2. *The product of two numbers with different signs is a negative number; and its module is equal to the product of the modules of the factors.*

3. *If one of the factors is equal to zero, then the product is equal to zero. On the contrary, the product is equal to zero, then and only when, at least one of the factors is equal to zero.*

4. *The operation with signs is this:*
   (a) $(+) \cdot (+) = +$;
   (b) $(+) \cdot (-) = -$;
   (c) $(-) \cdot (+) = -$;
   (d) $(-) \cdot (-) = +$.

3.9   Division of Rational Numbers

Division of rational numbers is considered as an operation which is opposite to that of multiplication. *We can define division of one number by another non-zero number as a process or operation of finding such a number which after multiplication by the divisor would yield the dividend as a product, that is:*

$$m \div n = q, \text{ if } n \cdot q = m; \ n \neq 0;$$

As discussed earlier, m is the *dividend*; n is the *divisor*, and q is the *quotient*. According to the above definition, a basic precondition of division is that the *divisor* should not, and cannot, be *zero*. Division of rational numbers by zero is avoided on account of the same reason or principle for which it is impossible to divide natural numbers by zero. There is no physical sense in dividing by zero. The result of dividing by zero is infinity (∞), which is *undefined* by reality. Having said that, except to operate with negative and positive signs, the division and multiplication of rational numbers is much the same in many respects as the division and multiplication of natural numbers.

Let us look at the following example problems below:

Problem 1.

(a) $(-17) \div (+2) = -8.5$;
(b) $(+12.369) \div (-3) = -4.123$;
(c) $(+45.14) \div (+7.4) = +6.1$;
(d) $(-18.56) \div (-3.2) = +5.8$.

Problem 2.

(a) $3x = -21.96$;
　　 $x = (-21.96) \div 3$;
　　 $x = -7.32$;

Proof:　 $3 \cdot (-7.32) = -21.96$;
　　　　 $-21.96 = -21.96$;

(b) $-12.5m = 25.25$;
　　 $m = 25.25 \div (-12.5)$;
　　 $m = -2.02$;

Proof:　 $-12.5 \cdot (-2.02) = 25.25$;
　　　　 $25.25 = 25.25$;

(c) $-5y = -35$;
　　 $y = (-35) \div (-5)$;
　　 $y = 7$;

Proof:　 $(-5) \cdot 7 = -35$;
　　　　 $-35 = -35$;

The operation with signs in the division of rational numbers is this:

(a) $(+) \div (+) = +$: The quotient of two numbers with the same signs is a positive number;

(b) $(+) \div (-) = -$: The quotient of two numbers with different signs is a negative number;

(c) $(-) \div (+) = -$: The quotient of two numbers with different signs is a negative number;

(d) $(-) \div (-) = +$; The quotient of two numbers with the same signs is a positive number;

In order to find the module of the quotient, it is necessary to divide the module of the dividend by the module of the divisor.

3.10 Circle and Circumference

Here we would like to draw a circle. For the radius of the circle we take the length of a line segment QR, as in Fig. 1.3 of section 1.8. We get a pair of *drawing compasses*; and we adjust its *spread* to be exactly equal to the length of segment QR. We then put the sharply pointed foot which is the needle of the compass to a point C somewhere on a sheet in your copybook to designate the center of the circle. Not changing the spread (or extent of opening) of the compass, we mark another point D (and also a few other points starting from D) by the second foot bearing the pencil around point C. All of the marked points including D are said to be equidistant from point C. If the line segment QR (also equal to CD) smoothly rotates around point C, then the second foot of the compass is said to have described an enclosed line, which is called a *circle*. The area A of a circle can be calculated using the formula $A = \pi \cdot r^2$, i.e.

$$\boxed{A_{circle} = \pi r^2}$$

where r is the radius of the circle. The measurement of the distance around the circle is called the *circumference* of the circle. Mathematicians have determined that the circumference of a circle is equal to the product of the diameter of the circle and a quantity known as *pi* (symbol π). That is, if we designate the letter c as the circumference of a circle and d as its diameter, then the circumference of the circle can be determined by the formula $c =$

πd. The value of π is equal to $22 \div 7 \approx 3.14$. The *diameter* of a circle is a straight line passing through the center of the circle and connecting two points on its surface. The *radius* **r** of a circle is equal to half of its diameter ($r = \frac{1}{2} d$).

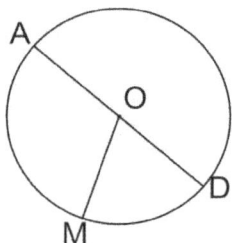

Fig. 3.2. Diameter and radius of a circle

In Fig. 3.2, line segment AD is the diameter of the given circle. Point O is the center of the circle. A line segment connecting the center of a circle and any arbitrary point on its surface is called the radius of that circle. The radius of a circle, by definition, is equal to half of its diameter. In Fig. 3.2, line segment OA = OD = OM is the radius of the given circle. The length of the radius of any given circle is a constant quantity, i.e. it does not change. This is because all the points on the surface (call it either the perimeter or circumstance) of the circle are equidistant from its center.

Also it should be noted that the center of a circle divides its diameter into two radii. The two points A and D, lying on the circumference (Fig. 3.2) divide the circle into two equal parts. Each part is called an *arch* of the circumference. Points A and D are the ends of these arches. The three radii OA, OD, and OM divide the circle into three parts, each of which is called a *sector*. The largest sector on the right is called a semicircle. A *semicircle* is one of the two equal parts into which the diameter divides a circle. Sector AOM is larger than sector MOD.

3.11   Areas of Rectangle and Triangle

The area of a square, the side of which is equal to a unit of length measurement, is usually accepted as the unit for measuring the areas of rectangles, triangles, and other similar geometric figures. For example, if the side of a square is one inch, then its area will be one square inch. Analogically, if the side of a square is one centimeter, then its area will be one square centimeter. A square centimeter can be written like this: cm^2. Small areas are measured in square millimeters (mm^2), square decimeters (dm^2), and square meters (m^2). See Fig. 3.3.

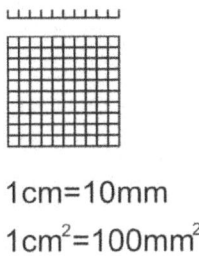

1cm=10mm

$1cm^2=100mm^2$

Fig.3.3. Areas of rectangle and triangle

For the measurement of areas of plots of land, they use such units as *are* (symbol a), hectare (symbol ha), and *square kilometer* (symbol km^2). An **are** is a unit of area equal to 100 square meters, or 119.599 square yards, or simply one hundredth of a hectare. A **hectare** is a unit of area equal to 100 ares, or to 10,000 square meters, or to 2.471 acres. An *acre* is a unit of area used in certain English-speaking countries, equal to 4840 square yards or 4046.86 square meters. A table is included in the appendix to this book which compares the basic units of measurement in the English and Metric systems. Another table is included which shows the relationship between the most prevalent units of area measurement.

In order to determine the area of a rectangle, it is necessary to find out how many units of area measurement it accommodates. If the sides of the rectangle are expressed in natural numbers w and l, then it is possible to slit the area of the rectangle into wl units of area measurement. See Fig. 3.4. Let us assume, for example, that the width of a rectangle w is equal to 4 cm and that the length l is equal to 7 cm. Then it is possible to slit this rectangle into 4 stripes in each of which will be 7 squares. Hence, the area A of the rectangle will be: A = 4 cm · 7 cm = 28 cm². The area of a rectangle, therefore, is determined by the formula $A = w \cdot l$, or

$$\boxed{A_{rectangle} = w \cdot l}$$

where w and l are the width and length, respectively, of the rectangle; and A is its area, measured in square units.

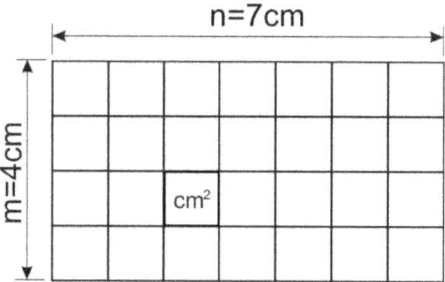

Fig. 3.4. Measurement of the area of a rectangle: A = 28 cm².

Let us designate the dimensions of a rectangle to be h and b, as shown in Fig. 3.5. Then the area A of the rectangle is calculated by the formula: $A = b \cdot h$. It is easy to see that the diagonal MN divides the rectangle into two equal triangles. This is why the area of each triangle is equal to half of the area of a rectangle, i.e. A = (b · h) ÷ 2, or

$$\boxed{A_{triangle} = \frac{1}{2}\,(bh)}$$

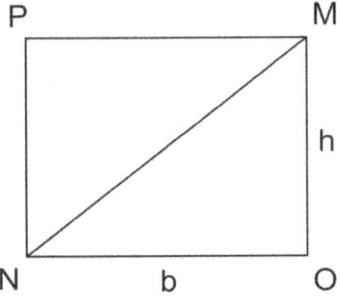

Fig. 3.5. The area of a triangle

Let us look at the triangle *MNO* in Fig. 3.6. We designate its base by the letter *b*. The perpendicular drawn from the vertex *M* to the base *NO* is said to be the *height* of the triangle. The length of line segment *MH* is the length of the height which we designate by the letter *h*. Then it is possible to determine the area *A* of this triangle by the formula:

$$A = \frac{1}{2} \ (bh)$$

where *b* is the length of the base, *h* – the height of the triangle drawn from its vertex to its base; and *A* is measured in square units.

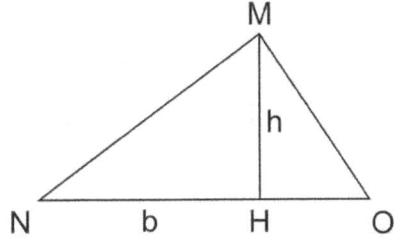

Fig. 3.6. The area of triangle MNO

3.12   Rectangular Parallelepiped

A *parallelepiped* (also called *parallelepipedon*) is a geometric figure or solid whose six sides or surfaces are parallelograms. A *parallelogram* is a quadrilateral or four-sided polygon whose opposite sides are parallel and equal in length. A parallelogram is a specific type of a rectangle. The forms of a book, pencil-box, box of matches, etc. are similar. They all give us an idea of a *rectangular parallelepiped.* Each side of a rectangular parallelepiped is a rectangle. As noted above, its opposite sides are equal to each other. The total surface of a rectangular parallelepiped consists of six rectangles, the area of which is equal to the sum of the areas of these rectangles. The sides of the rectangles, which are the borders of the parallelepiped, are called the ribs or edges of the rectangular parallelepiped. See Fig. 3.7.

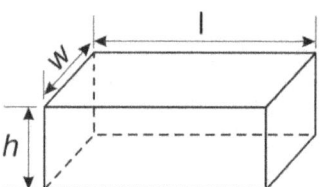

Fig. 3.7. A rectangular parallelepiped

A rectangular parallelepiped contains 12 edges, each 4 of which are equal to each other. Each rectangular parallelepiped has a *length, width*, and *height*, which are called its dimensions or measurements. *They are equal to the lengths of the three edges drawn from one vertex. Fig. 3.8 represents a rectangular parallelepiped, dimensions of which are equal to each other; i.e. the length, the width, and the height are of the same measurement. Such a rectangular parallelepiped is called a* **cube***.* The total surface of a cube comprises of six squares (see Fig. 3.8). The area of each square is equal to n^2, where n is equal to the length of the three edges drawn from a vertex. Therefore, the total surface of a cube represents an area A which is equal to $6 \cdot n^2$, i.e.

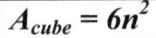

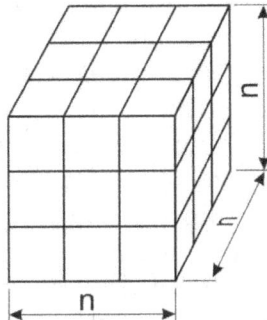

Fig. 3.8. A cube is also a rectangular parallelepiped

We can divide a cube into unit smaller cubes which will be equal to n^3. Draw a figure similar to Fig. 3.8 and show how one cubic decimeter (1 dm^3) can be divided into smaller unit cubic centimeters (cm^3). If you count them correctly, you will discover that there are 1000 cm^3 in one dm^3, i.e. 1 dm$^3 = 1000$ cm^3.

3.13   Volume of a Rectangular Parallelepiped

If a liquid is poured into some container, it may fill all of the space of this container which is determined by its cubic content. Then we can say that the container has a volume. *Volume* is a geometric quantity which characterizes the magnitude of the three-dimensional space enclosed within a given object. It is the capacity occupied by the object. Let us pour the liquid into another container. It may happen that the liquid won't completely fill the new container; or that all the liquid cannot be taken in. You can be convinced about this yourself. In this instance, we can say that the containers have different volumes since they can accommodate different quantity of liquid; or we can say that the containers are equal in volume, if the same quantity of liquid can be poured into them.

For a unit of volume it is accepted to use the volume of a cube, an edge of which is equal to a unit of length. *To calculate the volume of an object or geometric figure means to determine or find out how many such smaller cubes the object or figure contains.* Some prevalent units of volume measurement in the Metric system are: one cubic millimeter (1 mm^3); one cubic centimeter (1 cm^3); one cubic decimeter (1 dm^3); one cubic meter (1 m^3); and one cubic kilometer (1 km^3). The relationships among these units are shown in a table in the appendix to this book. In some English-speaking countries like Ghana, Gambia, Liberia, Nigeria, and the United states, the British units of volume measurement are usually used. Some prevalent units of volume measurement in the British system are: one cubic inch (1 inch^3); one cubic foot (1 ft.^{3}); one cubic yard (1 yard^3); and one cubic mile (1 mile^3).

In practice, volumes are calculated by definite formulas. In order to calculate a volume, the most important dimensions of an object or figure are its length, width, and height. Once these values are known, it is possible to calculate the volume V by the formula $V = l \cdot w \cdot h$, or

$$\boxed{V = l \cdot w \cdot h}$$

where l is the length, w is the width, h is the height, and *V is measured in cubic units.*

Consequently, the volume of a rectangular parallelepiped is equal to the product its dimensions, i.e. its length, weight, and height.

Let us take, for example, the rectangular parallelepiped in Fig. 3.7. If we arbitrarily designate its length (*l*) to be equal to 9 cm, width (*w*) to be equal to 5 cm, and height (*h*) to be equal to 6 cm, then we can determine its volume its volume using the formula given above:

$$V = l \cdot w \cdot h = 9 \ cm \cdot 5 \ cm \cdot 6 \ cm = 270 \ cm^3.$$

This means that a rectangular parallelepiped with the given measurements contains two hundred seventy cubic centimeters (270 cm^3). In other words, it contains 270 smaller cubes, each of which has its dimensions of length, width, and height equal to one cubic centimeter (1 cm^3). In essence, this confirms what it means to calculate a volume: that is, to find out how many smaller cubes an object or figure contains. What other example can you think about?

Let us look at another example in which the dimensions or edges of a rectangular parallelepiped are equal, as represented in Fig. 3.8. This means that its length, width, and height are equal, i.e. $l = w = h = n = a$. As we already know, such a rectangular parallelepiped is called a cube. In this case, the formula above becomes: $V = n \cdot n \cdot n$, or simply

$$\boxed{V_{cube} = n^3 = a^3}$$

The above formula tells us that the volume of a cube is equal to the cube of its edges.

Chapter Three: Questions to Test Your Understanding

1. What is the formula for determining the volume of a rectangular parallelepiped?

2. Name at least five units in which the volumes of geometric figures or objects are measured.

3. Explain what is meant by the volume of a geometric figure or object.

4. What is a cube?

5. What are the dimensions of a rectangular parallelepiped?

6. What are examples of objects that have the form of a rectangular parallelepiped?

7. What are the formulae for determining the following?
   (a) the area of a circle;
   (b) the area of a triangle;
   (c) the area of a rectangle;
   (d) the area of a cube;
   (e) the volume of a rectangular parallelepiped;
   (f) the volume of a cube;
   (g) the area of a square.

8. What are the respective units of measurements in question number 7?

9. Define the following terms: (a) circle; (b) circumference;
   (c) semicircle; (d) diameter; (e) radius.

10. Draw a circle and show its circumference, radius, and diameter.

11. Define the following terms: (a) multiplicand; (b) multiplier;
    (c) product; (d) factor; (e) dividend; (f)   divisor; (g) quotient.

12. How do we multiply natural numbers by ranked units such
    as 10, 100, 1000, and so on. Give examples.

13. How do we divide natural numbers by ranked units such as 10,
    100, 1000, and so on. Give examples.

14. Define the following:
    (a) Commutative property of multiplication
    (b) Associative property of multiplication
    (c) Distributive property of multiplication

Chapter Three: Problems and Exercises

1. Represent the sum in the form of a product:

(a) 534 + 534 + 534 + 534;
(b) 23 + 23 + ... (6 times);
(c) n + n + n + n;
(d) k + k + k + ... (n times).

2. Write in the form of a product:
   (a) the sum of five addends, each of which is equal to 5;
   (b) the sum of three addends, each of which is equal to n;
   (c) the sum of nine addends, each of which is equal to 1.

3. Name any two arbitrary natural numbers, product of which is equal to:
   (a) 11; (b) 0; (c) 15; (d) 20.

4. A motorist drives 64 miles in one hour. Calculate the distance in miles he will travel in: (a) n hours; (b) 4 hours; (c) 5 hours; (d) 7 hours.

5. Determine the unknown number x which satisfies each of the following equation, and check whether each equation is correct.
   (a) $(x + 2)(x + 1) = 6$;
   (b) $(x - 5)(x - 3) = 0$;
   (c) $(x + 3)(x - 3) = 7$;
   (d) $(x - 2)(x - 1) = 2$.

6. Check whether or not the division is correctly performed:
   (a) (i) $1,915 \div 1,915 = 0$;
       (ii) $44,424 \div 36 = 1,234$;
       (iii) $509 \div 1 = 509$;
       (iv) $1 \div 956 = 956$;

   (b) (i) $0 \div 1,214 = 0$;
       (ii) $2,189 \div 0 = 0$;
       (iii) $34,840 \div 65 = 536$;
       (iv) $47,035 \div 23 = 5,402$.

7. Solve the following problems:

(a) If four cups of rice cost 80 United States cents, for example, what is the cost of ten cups of rice?

(b) A store sells 5 yards of school uniform cloth for $20.75. At this rate how many yards of cloth can be bought for $62.25?

(c) A bicyclist traveled 60 miles within 3 hours; and a pedestrian for 5 hours walked 30 miles. By how many times the speed of the bicyclist was more than that of the pedestrian?

8. Calculate the followings by any convenient methods:
   (a) (i) $36 \cdot 50$;
   (ii) $60 \cdot 25$;
   (iii) $72 \cdot 250$;
   (iv) $342 \cdot 5$;
   (v) $144 \cdot 25$.
   (b) (i) $18,900 \div 900$;
   (ii) $90 \div 6 \cdot 400$;
   (iii) $25,350 \div 650$;
   (iv) $72,000 \div 400$;
   (v) $900 \div 450$.

9. Write the followings in:
   (a) kilograms:
       (i) 2 tons;
       (ii) 3 centners;
       (iii) 450 grams;
   (b) tons:
       (i) 450 centners,
       (ii) 5,000 kilograms;
   (c) centners:
       (i) n kilograms;
       (ii) k centners;
       (iii) a tons.

10. Find the number which is:
    (a) 100 times greater than 100;
    (b) 500 times less than 1,500.

11. A motorist drove with a speed of 75 mph for 5 hours. For which length of time it is possible to travel this distance on a bicycle with the speed of 25 mph?

12. Fifteen identical crates of soft drinks were brought to a bar. Each crate contained 9 rows with 6 bottles in a row. How many bottles of soft drinks were brought to the bar? Make up an expression for determining the quantity of soft drink bottles, if on the next day they brought x number of such crates.

13. Simplify the expressions: (a) $5a \cdot 6$; (b) $7k \cdot 3n \cdot 4$; (c) $4x \cdot 2 \cdot 5x$.

14. The dimensions of separate quarters of the premises of a sports complex are given in Table 3.1. Calculate the total area of the sports complex.

Table 3.1

№	Quarters	Width, yard	Length, yard	Area, yard²
i	Football field	75	120	
ii	Kitchen	4	6	
iii	Canteen	15	15	
iv	Service room	4	7	
v	Swimming pool	40	50	
vi	Administrative room	9	10	

15. How many rectangles with area 24 yards² are possible to form with the condition that the dimensions of their sides are expressed in whole numbers? Which rectangle has the least perimeter?

16. Calculate orally the following expressions:
   (a) $7 \cdot 11 \cdot 17 \cdot 0 \cdot 25$;
   (b) $2 \cdot 10 \cdot 5 \cdot 16 \cdot 4$;
   (c) $75 \cdot 124 \cdot 54 \cdot (21 - 21)$;
   (d) $5 \cdot 11 \cdot 2$.

17. Calculate the following by any convenient method:

(a) $125 \cdot 2 \cdot 486 \cdot 4$;

(b) $4 \cdot 735 \cdot 250$;

(c) $4 \cdot 567 \cdot 25$;

(d) $354 \cdot 8 \cdot 40 \cdot 125 \cdot 625$.

18. Calculate the expressions by two methods and tell which of them is simpler.

    (a) $(28 + 12) \cdot 15$;

    (b) $(18 + 32) \cdot 12$;

    (c) $(68 - 28) \cdot 13$;

    (d) $(112 - 72) \cdot 16$.

19. Two buses simultaneously left a car parking station and traveled in the same direction. One of them was driven at the speed of 45 mph, and the other at 60 mph. Which distance was between them after 3 hours?

20. Two buses simultaneously left a car parking station and traveled in opposite directions. One of them was driven at the speed of 75 mph, and the other at 68 mph. What was the distance between them after 4 hours?

21. Solve the following problems:

    (a) In the down stair accommodation of a cinema, there are 39 rows of sitting capacity with 23 seats in each row. In the upstairs (balcony) space, there are 18 rows with 23 seats in each row. What is the total number of seats in the cinema?

    (b) A charity organization "Vita" bought for an orphanage 105 textbooks at the price of $5.76 for each, and 160 chairs at $9.75 for each. What was the total amount for the two purchases by "Vita"?

    (c) How much more money was spent for the chairs than for the textbooks?

22. Open the parentheses or brackets and perform the indicated operations:

    (a) $6 \cdot (12 + 18)$;

    (b) $7 \cdot (15 - 9)$;

    (c) $15 \cdot (4 + a)$;

(d) $9 \cdot (m - 5)$;

(e) $3 \cdot (m + n)$;

(f) $7 \cdot (m + n - a)$

23. From two bus stations two buses started at the same time to travel in opposite directions towards each other and met after 3 hours. The speed of one was 53 mph, and that of the other was 65 mph. What was the distance between the two bus stations? Illustrate your solution by means of a drawing.

24. Calculate the followings:

(a) $49 \cdot 75,863$;

(b) $123 \cdot 45,678$;

(c) $9,786 \cdot 543$;

(d) $6,090 \cdot 705$.

25. Calculate the value of each expression:

(a) $506x + 32$, if: (i) $x = 25$; (ii) $x = 15$.

(b) $978a - 504$, if: (i) $a = 17$; (ii) $a = 12$.

26. Two steamers departed at the same time from two sea ports in opposite directions to meet each other. What is the distance between the two sea ports if the two steamers met after 6 hours. The average velocity of one of them is 42 mph and that of the other is 58 mph.

27. A motor boat sailed in a lake for two hours with the speed of 27 mph and later in a river for five hours. Calculate the distance it traveled for the 7 hours, if the speed of flow of the river is 4 mph. Examine possible cases of the motor boat moving downstream or upstream the river.

28. Find the quotient and remainder:

(a) $3,589,369 \div 769$;

(b) $975,864 \div 520$.

29. At which value of each letter will the equality be true?

(a) $580 \div x = 2$;

(b) $214 \div m = 214$;

(c) $0 \div a = 0$;

(d) $2 \div n = 1$.

30. Simplify the expressions:
    (a) $60x \div 12x$;
    (b) $25m \div m$;
    (c) $72n \div 12$;
    (d) $90kg \div 15kg$;
    (e) $a \div 1$.

31. Solve the equations:
    (a) $x \div 5 = 21$;
    (b) $2x + 8 = 20$;
    (c) $61 - x = 41$;
    (d) $(x + 11) \div 3 = 10$.

32. Solve the equations:
    (a) $51 \div (7 - x) = 17$;
    (b) $x \div 15 - 19 = 31$;
    (c) $75 \div ((x + 20) + 12 = 15$;
    (d) $x - 18 = 35$.

33. Fifteen textbooks were taken from a box containing x number of textbooks. The rest of the textbooks were then transferred into 18 cases. After the transfer of the rest of the textbooks into the 18 cases, it was found out that each of the 18 cases contained 24 textbooks. How many textbooks were there in the box before the transfer?

34. Calculate by any convenient method:
    (a) $(75 \cdot 15) \div (25 \cdot 5)$;
    (b) $(81 \cdot 49) \div (27 \cdot 7)$;
    (c) $(105 \cdot 90) \div (35 \cdot 45)$;
    (d) $(51 \cdot 44) \div (17 \cdot 22)$.

35. A father is 25 years older than his son. How old are the father and his son, if the son is 6 times younger than his father?

36. The sum of 2 numbers is 81. The first of them is 8 times greater than the second. Find the numbers.

37. Perform the indicated operations:
    (a) $38,581 + (12,724 - 133,945 \div 215)$;
    (b) $901,827 - (79,516 \div 412 + 160,062 \div 518)$;
    (c) $52,346 + 295,740 \div 372 - 136 \cdot 19$.

38. In dividing the number k by 15, we obtained the remainder to be two times greater than the quotient. What is the lowest value of the number k, which is a natural number.

39. Kojo thought of a certain number, added **23** to it, multiplied the sum by **5,** and subtracted from the product a number five times greater than the number he initially thought of. What result did Kojo obtain? Explain why the result does not depend on the number Kojo initially thought of.

40. Make up an expression for the solution of each problem below and find the solution to the problem.
    (a) Tundy walked 9 miles and rode 4 hours on his bicycle. With what speed did he ride his bicycle if all the distance he traveled was 45 miles? Graphically illustrate your solution to the problem.
    (b) Several textbooks were bought at $6 each and one textbook for $14. The total amount of $62 was paid for all the purchase. How many textbooks were bought in all?

41. Solve the indicated equations:
    (a) $36,225 \div x = 345$;
    (b) $36,225 \div (x - 475) = 345$.

42. Simplify the expressions:
    (a) $48 \cdot x \cdot 84$;
    (b) $y \cdot 23 \cdot 11$.

43. Two cars, the distance between them 250 miles, left at the same time towards each other and met after 5 hours. Find the speed of the second car, if the speed of the first was 30 mph.

44. Calculate the value of the given expression: $201 - (856 - 256) \div 3$.

45. What is the product of the roots of the equation: $(x - 7)(x - 15) = 0$?

46. In a three-room apartment, the first room is larger than the second by $5m^2$, and the second is larger than the third by $3m^2$. Find the total area of the 3 rooms, if the area of the smallest of them is $12m^2$.

47. How does the product **208 · 160** change, if the first factor is decreased 3 times and the second is increased 15 times?

48. What is the value of the expression:
$351 \cdot 45 - (127{,}500 + 63{,}750) \div 25$?

49. In the 5th grade class there are 9 pupils more than those in the 6th grade; and in the 6th grade class, there are 3 pupils more than those in the 4th grade. How many pupils in all three classes, if there are 31 pupils in the 4th grade?

50. How does the quotient of $510 \div 170$ change, if the dividend is increased three times, and the divisor is decreased ten times?

51. Draw a circle with a radius of 1.5 cm. Mark a point K inside of this circle. Plot two points on the surface of the circle distances from which to point K are the least and greatest, respectively. Compare these distances with the length of the radius.

52. Draw a circle with point P as the center and the radius is equal to 2cm 5mm. Mark a point S so that PS is equal to 4cm 5mm. With the help of a pair of compasses, find on the surface of the circle points

remote from point S by 4cm; by 8cm 5mm; 5cm; by 7cm; and by 1cm 6mm.

53. The least and greatest distances from point P (on a straight line passing through the center C of a circle), lying within the circle to the circumference respectively, are equal to 20 mm and 40 mm. Find the lengths of the radius and diameter of the circle.

54. Draw a circle with C as the center and radius is equal to 2 cm. Designate on its surface (circumference) point M; draw another circle with point M as the center and radius is equal to 2 cm. Will the circumference or surface of the second circle pass through point C? Explain.

55. Draw a circle with center C and radius 4 cm. The smallest distance from point B, which lies within the circle, to the surface is equal to 9 mm. Point D moves and or revolves around the surface (circumference). Find the greatest distance between points D and B.

56. Construct in the form of a linear diagram the lengths of the following rivers: Nile (6,741 km); Congo (4,800 km); Niger (4,184 km); Zambezi (2,740 km); Limpopo (1,770 km); and Volta (1,600 km).

57. Draw a circular diagram of the distribution of time spent by a pupil in the course of a day: sleeping – 8 hours; sports – 3 hours, entertainment and rest – 3 hours.

58. The following equalities are true for two given triangles: AB = LM; BC = MN; AC = LN; angle ABC = angle LMN; angle BCA = angle MNL; angle BAC = angle MLN. Represent the described equal triangles and make schematic drawings of them.

59. Draw an arbitrary rectangle and divide it into two different triangles by means of a diagonal joining any two given angles. To which kind of triangle do these triangles relate?

60. Point T lies within a rectangle MNOP. The distance of T from one of the shorter sides is 9 cm and from the other shorter side is 6 cm. Similarly, its distance from one longer side is 5 cm, and 3 cm from the other. Find the perimeter and area of this rectangle.

61. The perimeter of a rectangle is equal to 36 cm, and one of its sides is 4 cm. Find the area of this rectangle.

62. The perimeter of a square is equal to 76 cm. Calculate its area.

63. The perimeter of a rectangle is 540 cm. Its length is 50 cm more than its width. Calculate the area of the described rectangle. You may draw a sketch of the rectangle to facilitate your answer.

64. The perimeter of a rectangle, the length and width of which are expressed in natural numbers, is equal to 18 cm. How many rectangles exist with such perimeter? Which of them has the smallest area?

65. The length of an auto salon is 15 meters and the width is 12 meters. In the salon 18 cars are on display for exhibition. What area of space can each car occupy in the given salon?

66. It is required to erect a fence of a rectangular form around a building. The width of the building is 80 m; and its length is 40 m more than its width. How many poles are needed for the fence if the distance required between them is 5 m.

67. The area of a rectangle is 72 cm^2. Calculate the area of a new rectangle that has a length which is six times less than the length of the given rectangle; and width is four times less than the width of the given rectangle.

68. One side of a rectangle having the length of 18 cm was increased by 4 cm; and its area   was accordingly increased by 20 cm^2. What was the area of this initial rectangle? What is the perimeter of the new rectangle? Show drawing of the solution to the problem.

69. A square having the perimeter of 96 cm was divided into four triangles by straight lines   joining opposite vertices. Calculate the area of each triangle so formed.

70. The base NO of the triangle in Fig. 3.6 is equal to 48 cm, and the height (h) is equal to 42 cm. What is the area of this triangle?

71. Calculate the base NO of the triangle in problem number 70, if the height (h) is 14 cm and the area is 35 cm^2?

72. Find the sum of the lengths of all the edges of a cube and the area of its total surface, if the length of each edge is 10 cm.

73. Calculate the area of the total surface of a box of matches if the dimensions of its edges (length, width, and height) are 5 cm, 3.5 cm, and 1.5 cm, respectively. Make up a formula for calculating the area of the total surface of such a rectangular parallelepiped.

74. Calculate the volume of a cube the edge of which is 9 cm.

75. A block of ice (in the form of a rectangular parallelepiped) that was being kept in a freezer had the dimensions of 12 dm, 15 dm, and 7 dm. What was the mass of the water before freezing, if it is known that one cubic decimeter (1 dm^3) of ice is formed from 900 grams of water?

76. A stone-cutting instrument cuts stones into blocks, which have the form of a rectangular parallelepiped with dimensions 35 cm, 20 cm, and 5 cm. What is the mass of a block of stone, if it is known that a mass of 1 dm^3 of stone weighs 2 kilograms?

CHAPTER FOUR: COMMON FRACTIONS

4.1   Fractions and Fractional Numbers

In this chapter we will be dealing with fractions and fractional numbers and operations with them. This is because in our daily life we encounter numerous experiences and situations which require some knowledge and understanding of fractions and fractional numbers and how to operate with them. Such knowledge and understanding is especially relevant when you are buying some things in the market, shop or store, where you need to purchase a fraction or part of what is being sold. For example, a kilogram of sweets costs 7 dollars and you need to buy only 300 grams; or a yard of school uniform cloth costs 23 dollars, and you need to buy only 3 and a half yards of that cloth. Many other similar situations abound where we need to purchase only a part of a whole, or determine the fractional part of one thing or another.

Dealing with fractions ultimately requires us to perform some division. Therefore, let us try to divide an orange equally between two persons. It is obvious that not one of them will get the whole orange, but rather a half of, it if it is divided equally. And if it is required to equally divide that orange among three persons, each of them will receive only one third of the orange. It is also obvious that four persons equally dividing that orange, each of them will be entitled to only one fourth of it. In our daily interactions, we most often have to do with the words "one half", "one third", and "one fourth". All of them designate or define some part of a whole quantity. It is also common likewise to speak of other parts such as "one fifth", "one sixth", "one tenth", "one twentieth", "one hundredth", and so on.

And what if it is required to equally divide not only one orange, but rather three oranges among four people? It is not difficult to guess that each of them will get one fourth of each of the three oranges. In other words, each of the four people will receive three fourths of the three oranges.

Expressions such as three fourths $(\frac{3}{4})$, one fourth $(\frac{1}{4})$, one half $(\frac{1}{2})$, one third $(\frac{1}{3})$, one fifth $(\frac{1}{5})$, one sixth $(\frac{1}{6})$, one tenth $(\frac{1}{10})$, and so on are called fractions. *A fraction is simply defined as a part of a whole.*

As indicated above, fractions are represented or expressed by two natural numbers divided by a horizontal line or bar, which is regarded as a sign of division. The fractional expressions above are some examples. Other examples are $\frac{2}{5}$, $\frac{5}{6}$, $\frac{3}{10}$, $\frac{5}{8}$, etc. The number written above the line is called the *numerator* of the fraction. The numerator of a fraction indicates how many of the equal parts have been taken away. The number written below the line is called the *denominator* of the fraction. The denominator indicates how many equal parts into which a whole unit has been divided.

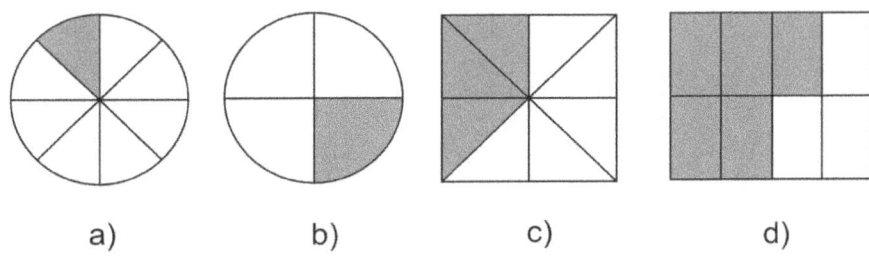

a)   b)   c)   d)

Fig. 4.1. Fractions of objects

In Fig.4.1 (a) a fraction $\frac{1}{8}$ is represented of which 1 is the numerator and 8 is the denominator. *You are required to write as fractions and read the remaining shaded and unshaded parts represented in (b), (c), and (d) of Fig.4.1 and explain how they were formed. Into how many equal parts is each figure divided? Which fractional part of each figure is shaded? Which fractional part of each figure is unshaded? Write these fractions in your copybook.*

Let us assume we have some coins with the value of 1 ¢, 2 ¢, 5 ¢, 10 ¢, 25 ¢, and 50 ¢. *Write in your copybook which fractional part of a dollar does each of these coins makes up.*

It is easy to observe from the above discussion that besides natural numbers there also exist *fractional numbers.* With their introduction, the operation of division has become possible in as much as it is feasible to regard a fraction as a quotient from the division of one number by another. Consequently, we are able to say that the *quotient* from the division of one natural number by another is a fraction, numerator of which can be regarded as the *dividend*; and the denominator can be regarded as the *divisor.* If, however, we divide one natural number by another and get a quotient as a result, then that quotient is said to be a fractional number *provided it is not a whole number.* The following examples show divisions in which the quotients or *results* are fractional numbers:

(a) $2 \div 3 = \dfrac{2}{3} = 0.666...;$

(b) $3 \div 5 = \dfrac{3}{5} = 0.6;$

(c) $1 \div 4 = \dfrac{1}{4} = 0.25;$

(d) $6 \div 10 = \dfrac{6}{10} = 0.6.$

Numbers of the form $\dfrac{a}{b}$, where a and b are natural numbers, are called *common fractions. In most common fractions, the denominator is usually greater than the numerator; that is, b is usually greater than a.* Notwithstanding, the opposite is possible. For example, $16 \div 2 = 8$; i.e. the quotient 8 is a natural number; or $\dfrac{16}{2} = 8$. It is possible to give a lot of such examples as: $\dfrac{12}{4} = 3$; $\dfrac{20}{5} = 4$; $\dfrac{28}{4} = 7$; and so on.

It follows from here, therefore, that it is possible to represent or write each natural number in the form of a fraction. Interestingly, each individual natural number can be written or expressed by different fractions. The simplest of all is to write a natural number in the form of a fraction, which has its denominator equal to 1; then in such case the numerator is equal to the given number, for example: $\frac{3}{1} = 3$; $\frac{5}{1} = 5$; $\frac{75}{1} = 75$. *In this context, it must be clearly understood and always remembered that the denominator of a fraction ca never be equal to zero.*

4.2   Proper and Improper Fractions

The numerator of a fraction may be less or greater than the denominator, or may be equal to it. Depending on the values of the numerator and denominator, fractions are divided into proper and improper fractions. A *proper fraction* is a fraction in which the numerator is less than the denominator; and an *improper fraction* is one in which the numerator is equal to or greater than the denominator. The fractions $\frac{2}{3}$, $\frac{4}{5}$, $\frac{3}{4}$, and $\frac{7}{10}$ are proper fractions; and $\frac{2}{2}$, $\frac{6}{5}$, $\frac{3}{1}$, and $\frac{10}{3}$ are improper fractions. Any improper fraction is possible to be represented in the form of a sum of a whole number and a proper fraction. For example, $\frac{10}{3} = 3 + \frac{1}{3} = 3\frac{1}{3}$.

Let us look at the solution to a simple problem:

Problem. Two boys climbed a coconut tree and picked 29 coconuts, which they divided equally between themselves. How many coconuts did each boy receive?

Solution. It is obvious that $29 \div 2 = \frac{29}{2}$ (coconuts). It is possible to reason otherwise: 29 (coconuts) = 28 (coconuts) + 1 (coconut); and dividing this into 2, we have 14 (coconuts) + $\frac{1}{2}$ (coconut). Therefore, each boy will receive 14 coconuts plus $\frac{1}{2}$ coconut. That is, $(14 + \frac{1}{2})$ coconuts. It is understood that $\frac{29}{2} = 14 + \frac{1}{2}$. In other words, the improper fraction $\frac{29}{2}$ is written in the form of the sum of a *whole part (natural number 14)* and a *fractional part (proper fraction $\frac{1}{2}$)*. A number consisting of a whole part and a fractional part is known as a *fractional number*. As you can see, a *fractional number* comprises of a whole number and a fraction; so it is also sometimes known as a *mixed number fraction*.

Let us look at another example. The height of Asatu is 160 centimeters. This is equal to $\frac{160}{100}$ meters, which can be written as 1 m 60 cm, or 1 m + $\frac{60}{100}$ m. Here the improper fraction $\frac{160}{100}$ is written in the form of the sum of a whole part and a fractional part $(1 + \frac{60}{100})$.

The improper fraction $\frac{36}{12}$ ($= 36 \div 12 = 3$) is equal a whole (natural) number without a fractional part. In this case, we say that its fractional part is equal to zero. Thus, it is possible to write any improper fraction in the form of the sum of a whole part and a fractional part. The whole part is a natural number, and the fractional component is either a proper fraction or zero.

If we remember *division with a remainder* in Section 3.6, then we can be reminded that the whole part of an improper fraction is an **incomplete quotient** as a result of dividing the numerator by the denominator of a fraction. In this case, the numerator of the fraction (fractional part) is the remainder in such given division.

Usually, the sum of the whole part and fractional part (i.e. the sum of the natural number and the fraction) is written as an entire number without a plus (+) sign. Consequently, instead of $3 + \frac{1}{2}$, it is customary to write $3\frac{1}{2}$, where 3 is the whole number, 2 is the denominator, and 1 is the numerator.

In order to convert an improper fraction into a *mixed number fraction*, it is necessary to divide the numerator by the denominator. The resulting incomplete quotient will be the whole number; the remainder of the division will be the numerator; and the denominator (or divisor) is always retained as the denominator of the derived mixed number fraction. For example:

(a) $\dfrac{160}{100} = 1 + \dfrac{60}{100} = 1\dfrac{60}{100}$;

(b) $\dfrac{17}{3} = 5 + \dfrac{2}{3} = 5\dfrac{2}{3}$;

(c) $\dfrac{29}{2} = 14 + \dfrac{1}{2} = 14\tfrac{1}{2}$;

(d) $\dfrac{31}{4} = 7 + \dfrac{3}{4} = 7\dfrac{3}{4}$.

An inverse operation is possible. It is easy to convert a fractional number (or a mixed number fraction) into an improper fraction. This is done *by multiplying the whole number by the denominator and adding the numerator to the product. This sum is the new numerator; and the original denominator is then retained as the denominator of the new improper fraction.* For example:

(a) $1\dfrac{60}{100} = \dfrac{160}{100}$;

(b) $5\dfrac{2}{3} = \dfrac{17}{3}$;

(c) $14\dfrac{1}{2} = \dfrac{29}{2}$;

(d) $7\dfrac{3}{4} = \dfrac{31}{4}$.

4.3   Comparison of Common Fractions

Let us compare $\frac{3}{10}$ m and $\frac{7}{10}$ m (see Fig. 4.2). Let 1 meter be divided into 10 equal parts. $\frac{3}{10}$ m designates that 3 such parts; and $\frac{7}{10}$ m designates the remaining 7 parts. It is obvious that $\frac{3}{10} < \frac{7}{10}$, since 3 < 7. It is read like this: "three tenths is less than seven tenths", or "seven tenths is greater than three tenths".

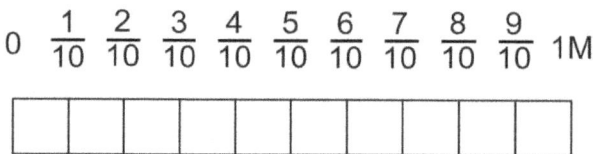

$$0 \quad \frac{1}{10} \quad \frac{2}{10} \quad \frac{3}{10} \quad \frac{4}{10} \quad \frac{5}{10} \quad \frac{6}{10} \quad \frac{7}{10} \quad \frac{8}{10} \quad \frac{9}{10} \quad 1M$$

Fig.4.2. Comparison of common fractions

RULE 1: *Of two given fractions with the same or equal denominators, the one having a greater (or less) numerator is greater (less). This means that in order to compare two fractions with identical denominators, it is necessary to compare their numerators.*

Let us compare the fractions $\frac{4}{15}$, $\frac{15}{15}$, and $\frac{18}{15}$. It follows from the preceding rule that $\frac{15}{15} > \frac{4}{15}$, since 15 > 4; and $\frac{18}{15} > \frac{15}{15}$, since 18 > 15. We are able to recall that $\frac{15}{15} = 15 \div 15 = 1$. It follows therefore that $\frac{4}{15} < 1$, and $\frac{18}{15} > 1$. In the same way, it is possible to affirm that $\frac{6}{6} = 1$; $\frac{5}{6} < 1$; and $\frac{7}{6} > 1$.

RULE 2: *If the numerator of a given fraction is equal to the denominator, then that fraction is equal to 1. If the numerator of a fraction is greater than (less than) the denominator, then that fraction is greater than (less than) 1.*

It is explicit from the preceding rule that a fraction is said to be *proper* if it is less than 1, and improper if it is either equal to or greater than 1.

In the discussions above, we have compared *fractions with identical denominators*. One question that springs to our mind is "How do we compare *fractions with different denominators*?" We shall find out later.In the mean time, you may think how you can briefly compare fractions with identical numerators such as $\frac{1}{7}$ and $\frac{1}{9}$, as well as $\frac{4}{7}$ and $\frac{4}{9}$.

RULE 3: *In two given fractions having identical numerators, the one whose denominator is greater (less) is less (greater).*

According to the preceding rule, $\frac{1}{7} > \frac{1}{9}$ (one seventh is greater than one ninth); and $\frac{4}{7} > \frac{4}{9}$ (four sevenths is greater than four ninths). *How can you convince and explain to someone that this is true? What other examples can you think about regarding the comparison of fractions with identical numerators but different denominators?*

RULE 4: *In comparing two given fractions having different numerators and denominators, it is necessary to convert them to their LCM (Least Common Multiple) and then determine their corresponding numerators.*

For example, how do we compare the fractions $\frac{2}{3}$ and $\frac{3}{4}$? Which of them is greater? We know that the LCD of 3 and 4 is 12; so we multiply $\frac{2}{3}$ by $\frac{4}{4}$ (that is, $\frac{2 \cdot 4}{3 \cdot 4}$) to get $\frac{8}{12}$; and we likewise multiply $\frac{3}{4}$ by $\frac{3}{3}$ (that is, $\frac{3 \cdot 3}{4 \cdot 3}$) to get $\frac{9}{12}$. Now we can easily compare the fractions

$\frac{8}{12}$ and $\frac{9}{12}$, having the same denominator; it is obvious that $\frac{9}{12}$ is greater than $\frac{8}{12}$. Consequently, $\frac{3}{4} > \frac{2}{3}$ or $\frac{2}{3} < \frac{3}{4}$.

Class work:

1. With the help of a pair of line segments, each pupil should compare the following fractions, using the signs either <, >, or = :

(a) $\frac{3}{5}$ and $\frac{5}{8}$;   (b) $\frac{7}{10}$ and $\frac{4}{5}$;   (c) $\frac{4}{9}$ and $\frac{5}{12}$;

(d) $\frac{3}{7}$ and $\frac{5}{9}$;   (e) $\frac{3}{4}$ and $\frac{7}{8}$

2. Given the following fractions: $\frac{3}{4}, \frac{7}{10}, \frac{3}{7}, \frac{3}{5}, \frac{4}{9}, \frac{5}{8}, \frac{5}{9}, \frac{5}{12}, \frac{4}{5}, \frac{7}{8}$.
   (a) Place them in increasing order.
   (b) Place them in decreasing order.

4.4   Basic Property of Fractions

Let us assume that the length of the line segment ST in Fig. 4.3 is 12 inches (12 inches is equal to 1 ft.). If we divide it into 8 equal parts with point R in the center, then SR = RT = 6 inches = $\frac{4}{8}$ (12 inches) as shown in Fig. 4.3 (a). If we divide line segment ST into 6 equal parts, then SR = RT = 6 inches = $\frac{3}{6}$ (12 inches) as in Fig. 4.3 (b). If ST is divided into 2 equal parts, then SR = RT = 6 inches = $\frac{1}{2}$ (12 inches) as in Fig. 4.3 (c).

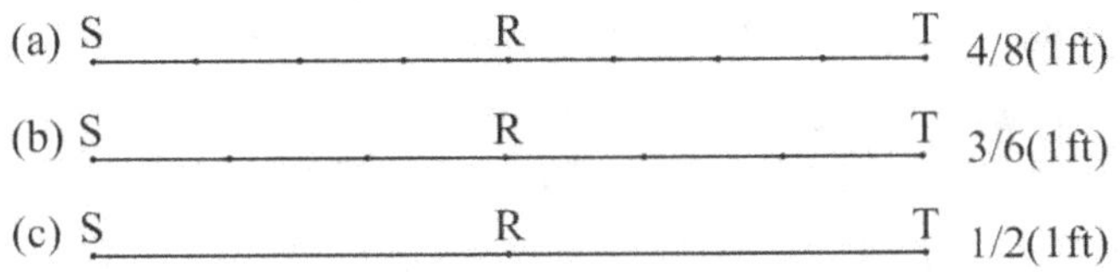

(a) S ———————————— R ———————————— T   4/8(1 ft)

(b) S ———————————— R ———————————— T   3/6(1 ft)

(c) S ———————————— R ———————————— T   1/2(1 ft)

Fig. 4.3. Basic property of fractions

The fractions $\frac{4}{8}$, $\frac{3}{6}$, and $\frac{1}{2}$ are different forms of writing the same and equal numerical value, that is:

$$\frac{4}{8} = \frac{3}{6} = \frac{1}{2}; \text{ or } \frac{1}{2} = \frac{3}{6} = \frac{4}{8}.$$

It is not difficult to see that:

$$\frac{1}{2} = \frac{1 \cdot 2}{2 \cdot 2} = \frac{1 \cdot 3}{2 \cdot 3} = \frac{1 \cdot 4}{2 \cdot 4},$$

or

$$\frac{4}{8} = \frac{4 \cdot 4}{8 \cdot 4} = \frac{4 \cdot 3}{8 \cdot 3} = \frac{4 \cdot 2}{8 \cdot 2}.$$

As noted earlier, the fractions $\frac{4}{8}$, $\frac{3}{6}$, and $\frac{1}{2}$ are distinguished only by the forms of their writing. They are in essence of the same value. The reason is explained in the following:

if the numerator and denominator of a fraction are multiplied or divided by the same natural number, then the result is a fraction equal to the given fraction.

This property is called the *basic property of fractions*. Division of the numerator and denominator of a fraction by their common divisor (different from 1) is called a *reduction of the given fraction*.

However, not every fraction may be reduced. For instance, the fraction $\frac{9}{16}$ cannot be reduced, inasmuch as its numerator and denominator, except 1, do not have common divisors. Just in the same way, fractions $\frac{2}{3}$, $\frac{4}{7}$, $\frac{5}{11}$, as well as $\frac{1}{3}$, $\frac{1}{10}$, $\frac{1}{50}$, that is, such fractions numerator of which is equal to 1, cannot be reduced. Fractions which are impossible to reduce are called *irreducible fractions* (or *simplified fractions*).

4.5   Reduction of Fractions to a Common Denominator

The operation we performed above in Rule 4 of section 4.3 (Comparison of Common Fractions) is called *reduction of fractions to a common denominator*. It is frequently necessary to change or convert fractions so that their denominators are equal. This is possible to do, using the basic property of fractions (section 4.4). That is, if we increase the denominator of a given fraction by a certain number of times, then in order that the value of that fraction is not changed, it is necessary to increase its numerator by as much times as the denominator has been increased. In this way, we can say that it is possible to *reduce* fractions with different denominators *to a common denominator*. The first step in *reducing fractions to a common denominator* is finding a common multiple of the denominators. For simplicity of calculations, it is expedient to find the *least common multiple*, which is usually referred to as the *LCM* of the denominators. (Note: the LCM or *least common multiple* is sometimes called LCD for *least common denominator* in certain text books.)

Problem: For example, if we were required to perform an addition operation over the fractions ($\frac{1}{2} + \frac{2}{3} + \frac{4}{5} + \frac{3}{10} + \frac{2}{15}$), the *least common multiple* of the denominators of the given fractions will be LCM (2, 3, 5, 10, 15) =
$= 2 \cdot 3 \cdot 5 = 30.$

Solution: We write the given fractions one under another; those equal by value are written in a column.

$$\frac{1}{2} \quad + \quad \frac{2}{3} \quad + \quad \frac{4}{5} \quad + \quad \frac{3}{10} \quad + \quad \frac{2}{15} \quad =$$

$$\frac{1\cdot 15}{2\cdot 15} + \frac{2\cdot 10}{3\cdot 10} + \frac{4\cdot 6}{5\cdot 6} + \frac{3\cdot 3}{10\cdot 3} + \frac{2\cdot 2}{15\cdot 2} =$$

$$\frac{15}{30} + \frac{20}{30} + \frac{24}{30} + \frac{9}{30} + \frac{4}{30} =$$

$$= \frac{72}{30} = \frac{12}{5} = 2\frac{2}{5}.$$

In order to reduce fractions to a common denominator, the following steps are necessary:

1. Find the lowest common multiple (LCM) of the denominators;
2. Determine a twiddle factor for each denominator;
3. Multiply the numerator and denominator of each fraction by a corresponding twiddle factor.

4.6 Addition and Subtraction of Fractions with Identical Denominators

Let us examine the following two problems:

Problem 1. A line segment with the length of 1 meter has been divided into 5 equal parts. See Fig.4.4 below. Making use of the drawing in the figure, the lengths of the line segments OK, KM, and OM are measured in meters. It is required to find the sum of the segments OK + KM.

Solution 1. The sum of $OM = OK + KM = \frac{1}{5}\,m + \frac{2}{5}\,m = \frac{3}{5}\,m.$

Conversely, KM = OM – OK .

The difference $KM = \frac{3}{5}\,m - \frac{1}{5}\,m = \frac{2}{5}\,m.$

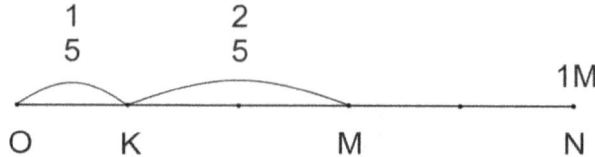

Fig. 4.4. Addition of fractions having identical denominators

Problem 2. Let us look at the drawing in Fig.4.5 shown below. It is required to determine the difference of the segments OC – OA and OC – AC.

Solution 2. The difference of $OC - OA = \frac{9}{6} - \frac{4}{6} = \frac{5}{6}$; and $OC - AC$
$= \frac{9}{6} - \frac{5}{6} = \frac{4}{6} = \frac{2}{3}$.

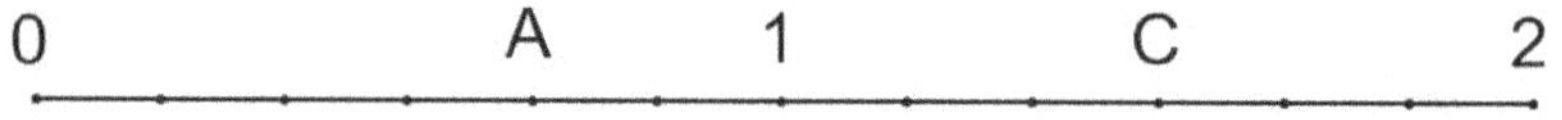

Fig.4.5. Subtraction of fractions having identical denominators

We can make the following conclusion regarding the addition and subtraction of common fractions: If the denominators of some given common fraction are identical, then we should add or subtract their numerators. That is, if we are required to add $\frac{1}{5}$ to $\frac{2}{5}$, then we will get $\frac{3}{5}$. And, conversely, if we subtract $\frac{1}{5}$ from $\frac{3}{5}$, then we will get $\frac{2}{5}$. The operations can be performed like this:

(a) $\frac{1}{5} + \frac{2}{5} = \frac{3}{5}$; and $\frac{3}{5} - \frac{1}{5} = \frac{2}{5}$;

(b) $\frac{9}{6} - \frac{4}{6} = \frac{5}{6}$; and $\frac{9}{6} - \frac{5}{6} = \frac{4}{6} = \frac{2}{3}$.

RULE 5: *In order to add (subtract) fractions with identical denominators, it is necessary to add (subtract) their numerators and retain as is. In a general case, we can write the formula like this:*

$$\frac{m}{k} \pm \frac{n}{k} = \frac{m \pm n}{k}$$

where ***m > n,*** **or** ***m = n; and k ≠ 0.***

Addition and subtraction of fractional numbers (so-called *mixed number fractions*) are performed with the help of the properties of operations as discussed. For example:

(a) $4\frac{2}{5} + 3\frac{1}{5} = (4 + \frac{2}{5}) + (3 + \frac{1}{5}) = (4 + 3) + (\frac{2}{5} + \frac{1}{5}) = 7 + \frac{3}{5} = 7\frac{3}{5};$

(b) $4\frac{2}{5} - 3\frac{1}{5} = (4 + \frac{2}{5}) - (3 + \frac{1}{5}) = (4 - 3) + (\frac{2}{5} - \frac{1}{5}) = 1 + \frac{1}{5} = 1\frac{1}{5};$

(c) $5\frac{4}{7} + 2\frac{2}{7} = (5 + \frac{4}{7}) + (2 + \frac{2}{7}) = (5 + 2) + (\frac{4}{7} + \frac{2}{7}) = 7 + \frac{6}{7} = 7\frac{6}{7};$

(d) $5\frac{4}{7} - 2\frac{2}{7} = (5 + \frac{4}{7}) - (2 + \frac{2}{7}) = (5 - 2) + (\frac{4}{7} - \frac{2}{7}) = 3 + \frac{2}{7} = 3\frac{2}{7};$

(e) $3\frac{5}{8} + 1\frac{2}{8} = (3 + \frac{5}{8}) + (1 + \frac{2}{8}) = (3 + 1) + (\frac{5}{8} + \frac{2}{8}) = 4 + \frac{7}{8} = 4\frac{7}{8};$

(f) $6 - 3\frac{1}{7} = 5\frac{7}{7} - 3\frac{1}{7} = (5 - 3) + (\frac{7}{7} - \frac{1}{7}) = 2 + \frac{6}{7} = 2\frac{6}{7}.$

RULE 6: *Where a whole number is given to be operated on with a common fraction and/or mixed number fraction such as in the case of (f) above, it is necessary first to convert the whole number into a mixed number fraction, using the denominator of the given fraction or fractions.* **For example:**

(a) $6 = 5\dfrac{7}{7} = 5\dfrac{1}{1} = 5\dfrac{4}{4} = \dots$ (and so on)

(b) $1 = \dfrac{3}{3} = \dfrac{5}{5} = \dfrac{2}{2} = \dots$

(c) $4 = 3\dfrac{6}{6} = 3\dfrac{7}{7} = 3\dfrac{10}{10} = \dots$

4.7 Addition and Subtraction of Fractions with Unlike Denominators

We had already added and subtracted fractions with identical denominators with the help of the following formulas:

$$\frac{m}{k} + \frac{n}{k} = \frac{m+n}{k} \;\; ; \;\; \frac{m}{k} - \frac{n}{k} = \frac{m-n}{k},$$

where $m > n$, or $m = n$; and $k \neq 0$. In cases where $m < n$, the approach is simply to add k to m (i.e., $k + m$) and then proceed as usual.

In order to add (or subtract) fractions with unlike or different denominators, it is necessary to reduce them to a least common denominator, carry out the addition or subtraction of their numerators (and also of their whole numbers in the case of *mixed number fractions*) and then use their least common denominator. The example and solution demonstrated in section 4.5 are typical of addition and subtraction of fractions with different denominators. Additional examples are given below:

1) $\dfrac{3}{4} + \dfrac{2}{5} + \dfrac{5}{8} = \dfrac{30}{40} + \dfrac{16}{40} + \dfrac{25}{40} = \dfrac{30+16+25}{40} = \dfrac{71}{40} = 1\dfrac{31}{40}$;

2) $\dfrac{2}{3} - \dfrac{7}{12} = \dfrac{8}{12} - \dfrac{7}{12} = \dfrac{8-7}{12} = \dfrac{1}{12}$;

3) $4\dfrac{7}{20} + 5\dfrac{1}{5} + 2\dfrac{1}{3} + 3\dfrac{1}{4} = 4\dfrac{21}{60} + 5\dfrac{12}{60} + 2\dfrac{20}{60} + 3\dfrac{15}{60} =$

$= (4 + 5 + 2 + 3) + \dfrac{21+12+20+15}{60} =$

$= 14 + \dfrac{68}{60} = 14 + 1 + \dfrac{8 \div 4}{60 \div 4} = 15 + \dfrac{2}{15} = 15\dfrac{2}{15}$;

4) $10\frac{3}{7} - 5\frac{7}{9} = 10\frac{27}{63} - 5\frac{49}{63} = (10-5) + \frac{27-49}{63} =$

$= 5 + \frac{27-49}{63} = 4 + \frac{(27+63)-49}{63} = 4 + \frac{41}{63} = 4\frac{41}{63};$

5) $6 - 2\frac{1}{2} = 5\frac{2}{2} - 2\frac{1}{2} = (5-2) + (\frac{2}{2} - \frac{1}{2}) = 3 + \frac{2-1}{2} = 3 + \frac{1}{2} = 3\frac{1}{2}.$

4.8   Conversion of Common Fractions to Decimal Fractions

In the solving of various problems, there are cases when both common and decimal fractions are involved. In such cases, the approach is to convert the common fraction(s) to decimal fraction(s) or vice versa.

Conversion of a common fraction to a decimal fraction (or simply decimal) can be conveniently carried out by dividing the numerator by the denominator. Examples are given below:

1) Convert the common fraction $\frac{1}{8}$ to a decimal fraction.

$$\frac{1}{8} = 0.125$$

$$
\begin{array}{r}
0.125 \\
8\overline{)1.000} \\
-\underline{8} \\
20 \\
-\underline{16} \\
40 \\
-\underline{40} \\
0
\end{array}
$$

2) Convert the common fraction $\frac{5}{16}$ to a decimal fraction.

$$\frac{5}{16} = 0.3125$$

$$
\begin{array}{r}
0.3125 \\
16\overline{)5.0000} \\
-\underline{48} \\
20 \\
-\underline{16} \\
40 \\
-\underline{32} \\
80 \\
-\underline{80} \\
0
\end{array}
$$

It is quite simple to convert a decimal fraction (or decimal) to a common fraction. To do this, simply write the corresponding decimal with a denominator. It is important to take note of the number of decimal places indicated by the decimal point in a decimal fraction. *To determine the denominator in a given case, just multiply the number of decimal places by ten.* Reduce the resulting fraction, if it is possible. See the examples below:

1) $0.5 = \frac{5}{10} = \frac{1}{2}$;

2) $0.02 = \frac{2}{100} = \frac{1}{50}$;

3) $0.371 = \frac{371}{1000}$;

4) $0.4309 = \frac{4309}{10000}$;

5) $6.25 = 6\frac{25}{100} = 6\frac{1}{4}$.

4.9   Decimal Approximations of Common Fractions

It is not always possible to convert a common fraction to an *exact* decimal fraction. For example, it is not possible to *exactly* convert the fraction $\frac{1}{6}$ to a decimal. The result of dividing the numerator by the denominator shows an *infinite* decimal fraction. Therefore, it is convenient only to write a decimal approximation of the common fraction, that is, to write in decimal notation an approximate value of the given fraction. See below, for example, the cases of converting the common fractions $\frac{1}{6}$ and $\frac{7}{12}$ to decimal fractions:

$$
6\overline{)\begin{array}{l}0.1666\\1.0000\end{array}} \quad ; \quad \frac{1}{6} = 0.1666...
$$

$$
\begin{array}{r}
-6 \\ \hline
40 \\
-36 \\ \hline
40 \\
-36 \\ \hline
40 \\
-36 \\ \hline
4
\end{array}
$$

$$
12\overline{)\begin{array}{l}0.58333\\7.00000\end{array}} \quad ; \quad \frac{7}{12} = 0.58333...
$$

$$
\begin{array}{r}
-60 \\ \hline
100 \\
-96 \\ \hline
40 \\
-36 \\ \hline
40 \\
-36 \\ \hline
40 \\
-36 \\ \hline
4
\end{array}
$$

The numbers 0.1666... and 0.58333... are not exact conversions to decimal of the common fractions $\frac{1}{6}$ and $\frac{7}{12}$. The three dots at the end of each of the numbers mean "and so on", since it is possible to infinitely continue the process of dividing. In such cases, we can talk about *approximate conversion* or *decimal approximations of common fractions*. For instance:

(a) $\dfrac{1}{6} \approx 0.1666... \approx 0.167 \approx 0.17$;

(b) $\dfrac{7}{12} \approx 0.58333... \approx 0.583 \approx 0.58$;

(c) $\dfrac{4}{7} \approx 0.571428571428571428... \approx 0.571428 \approx 0.57$;

(d) $\dfrac{15}{37} \approx 0.405405405... \approx 0.405 \approx 0.4$;

(e) $\dfrac{3}{11} \approx 0.272727... \approx 0.272 \approx 0.27$.

Infinite decimal fractions in which one or several digits are repeated in the same sequence are called *periodic decimal fractions (or recurrent decimal fractions)*. Periodicity of decimal fractions may be either *pure* or *mixed*. *Purely periodic decimal fractions are those in which the period begins at once after the decimal point.* Examples of purely periodic decimal fractions are 0.571428571428571428..., 0.405405405..., 0.343434..., etc. A *mixed periodic decimal fraction is such decimal in which there is one or several non-recurring (non-repeating) digits between the decimal point and the first period.* Examples of mixed periodic decimal fractions are 0.1666..., 0.58333..., 0.231565656..., etc.

It is customary to write a periodic decimal fraction in an *abbreviated form*. For example, instead of 0.571428571428571428... it is accepted to write 0.(571428); instead of 0.405405405... they write 0.(405); that is, the period is written in a bracket. Other examples are:

(a) **0.343434... = 0.(34) – decimal fraction with period 34;**
(b) **0.1666... = 0.1(6) – decimal fraction with period 6;**
(c) **0.58333... = 0.58(3) – decimal fraction with period 3;**
(d) **0.231565656... = 0.231(56) – decimal fraction with period 56.**

4.10   More about Common Fractions

4.10.1   Mutually Inverse Numbers

At times we notice that *the product of two fractions is equal to 1. This is possible if, and only when, the numerator of the first fraction is the denominator of the second, and the denominator of the first fraction is numerator of the second.* Thus, numbers of the form $\frac{m}{n}$ and $\frac{n}{m}$ are called *mutually inverse (or mutually reciprocal) numbers. As a rule, the product of mutually inverse numbers such as* $\frac{m}{n}$ *and* $\frac{n}{m}$ *is equal to 1:*

$$\frac{m}{n} \cdot \frac{n}{m} = 1.$$

Example Problem 1. **Instead of the letters, supply the necessary numbers which will make the equations true:**

(a) $\frac{5}{17}y = 1;$ (b) $\frac{8}{11}x = 1;$ $2\frac{2}{3}d = 1.$

Example Problem 2.

(a) **By which fraction is it possible to multiply** $1\frac{1}{2}$ **so as to get 1 as the product?**

(b) **Find the products of (i)** $\frac{5}{7} \cdot \frac{7}{5};$ **(ii)** $\frac{3}{4} \cdot \frac{4}{3}.$

4.10.2   Multiplication of Common Fractions

Let us go back to the rectangle in Fig. 3.4 (section 3.11, Areas of Rectangle and Triangle). Suppose we have a problem in which we are required to find not the entire area of the rectangle in Fig. 3.4, but rather only a part of it. In particular, we are interested in finding the area of $\frac{3}{4}$ of the width and $\frac{6}{7}$ of the length.

We can see from the drawing in Fig. 3.4 that, from one side, the length of the rectangle is divided into 7 equal parts; and, from the other side, the width is divided into 4 equal parts. The given area of the rectangle contains 28 smaller rectangles. Every side of each smaller rectangle is equal to one centimeter. This means that each smaller rectangle is essentially a square centimeter, i.e. 1 cm^2. Therefore, we can say that the area of the given rectangle contains 28 square centimeters (or is equal to 28 cm^2); and one such smaller rectangle makes up $\frac{1}{28}$ part of the entire area of the rectangle. The required area of the part of the rectangle in which we are interested ($\frac{6}{7}$ of the length and $\frac{3}{4}$ of the width) contains 18 of such parts ($\frac{6}{7} \cdot \frac{3}{4}$). Consequently, we can say that the product of two numbers $\frac{6}{7}$ and $\frac{3}{4}$ is the number $\frac{18}{28} = \frac{9}{14}$. So the solution to our problem will be:

$$A = l \cdot w = \frac{6}{7} \cdot \frac{3}{4} = \frac{18}{28} = \frac{9}{14} \ (\text{cm}^2).$$

The product of two fractions is equal to a fraction, the numerator of which is equal to the product of their numerators, and the denominator is equal to the product of their denominators, as formulated below:

$$\frac{m}{n} \cdot \frac{x}{y} = \frac{m \cdot x}{n \cdot y}.$$

The above formula can be used for the multiplication of both natural numbers and fractions. To do this, it is necessary to write a natural number in the form of a fraction and then apply the above formula. Some examples are given below:

$$\textbf{(a)}\ 7 \cdot \frac{3}{5} = \frac{7}{1} \cdot \frac{3}{5} = \frac{21}{5} = 4\frac{1}{5};$$

$$\textbf{(b)}\ \frac{3}{4} \cdot 12 = \frac{3}{4} \cdot \frac{12}{1} = \frac{36}{4} = \frac{9}{1} = 9;$$

(c) $\frac{4}{5} \cdot 25 = \frac{4 \cdot 25}{5} = \frac{4 \cdot 5}{1} = \frac{20}{1} = 20;$

(d) $\frac{2}{3} \cdot \frac{9}{11} = \frac{2}{1} \cdot \frac{3}{11} = \frac{6}{11}.$

Multiplication of fractions obeys the commutative and associative properties:

(a) $\frac{5}{8} \cdot \frac{7}{10} = \frac{7}{10} \cdot \frac{5}{8} = \frac{35}{80} = \frac{7}{16}$ - commutative property of multiplication;

(b) $(\frac{2}{3} \cdot \frac{4}{5}) \cdot \frac{6}{7} = \frac{2}{3} \cdot (\frac{4}{5} \cdot \frac{6}{7})$ – associative property of multiplication.

Just as with the help of decimal fraction, a lot of practical problems can likewise be solved with the help of common fractions.

Example problem. Yango went into a supermarket and spent the amount of $51.00 on the purchase of certain items in the supermarket. The $51.00 is $\frac{3}{17}$ of all the money Yango had carried into the supermarket. How much money did he carry into the supermarket?

Solution. In order to solve this problem, it is necessary to find $\frac{3}{17}$ of *what number* is equal to 51, i.e. $\frac{3}{17} \cdot x = 51$, where the *unknown number* represented by x can be found by multiplying 51 by $\frac{17}{3}$:

$x = \frac{17}{3} \cdot 51 = \frac{17 \cdot 51}{3} = \frac{17 \cdot 17}{1} = \frac{289}{1} = \$289$ – the amount of money Yango carried into the shop.

4.10.3   Division of Common Fractions

Problem. We are required to find the length of a rectangle having an area of $\frac{19}{25}$ cm^2; and its width is $\frac{4}{15}$ cm.

Solution. We can make up an equation with one variable. Let's assume that the required length of the rectangle is x (cm). Then the area (A) of the rectangle would be equal to $A = l \cdot w = w \cdot l = \frac{4}{15} \cdot x$ (cm^2). Consequently:

$$\frac{4}{15} \cdot x = \frac{19}{25}.$$

It follows from here that:

$$x = \frac{19}{25} \div \frac{4}{15}.$$

Since we do not yet know how to divide one fraction by another, we solve the problem otherwise. We multiply both sides of the equation $\frac{4}{15} \cdot x = \frac{19}{25}$ by a number inverse (or reciprocal) to the factor $\frac{4}{15}$, that is, by $\frac{15}{4}$:

$$\left(\frac{4}{15} \cdot x\right) \cdot \frac{15}{4} = \frac{19}{25} \cdot \frac{15}{4},$$

or

$$\left(\frac{4}{15} \cdot \frac{15}{4}\right) \cdot x = \frac{19}{25} \cdot \frac{15}{4};$$

$$1 \cdot x = \frac{19}{25} \cdot \frac{15}{4} \rightarrow x = \frac{19 \cdot 15}{25 \cdot 4} = \frac{19 \cdot 3}{5 \cdot 4} = \frac{57}{20} \rightarrow x = 2\frac{17}{20} \text{ (cm)}.$$

We have found the number x, such that $\frac{4}{15} \cdot x = \frac{19}{25}$. This number is a *quotient* from the division of $\frac{19}{25}$ by $\frac{4}{15}$. In this way,

$$\frac{19}{25} \div \frac{4}{15} = \frac{19}{25} \cdot \frac{15}{4}.$$

Therefore, we can formulate the following rule: *in order to divide one fraction by another, it is necessary to multiply the dividend by a number inverse (or reciprocal) to the divisor. In other words, it is necessary to invert the divisor; and then multiply the dividend by the inverted divisor:*

$$\frac{m}{n} \div \frac{x}{y} = \frac{m}{n} \cdot \frac{y}{x}.$$

Some examples are given below:

(a) $\frac{5}{6} \div \frac{3}{8} = \frac{5}{6} \cdot \frac{8}{3} = \frac{5 \cdot 8}{6 \cdot 3} = \frac{5 \cdot 4}{3 \cdot 3} = \frac{20}{9} = 2\frac{2}{9};$

(b) $\frac{1}{2} \div \frac{5}{16} = \frac{1}{2} \cdot \frac{16}{5} = \frac{1 \cdot 8}{1 \cdot 5} = \frac{8}{5} = 1\frac{3}{5};$

(c) $3\frac{3}{4} \div \frac{9}{10} = \frac{15}{4} \cdot \frac{10}{9} = \frac{5 \cdot 5}{2 \cdot 3} = \frac{25}{6} = 4\frac{1}{6};$

(d) $\frac{4}{5} \div 2\frac{1}{2} = \frac{4}{5} \div \frac{5}{2} \to \frac{4}{5} \cdot \frac{2}{5} = \frac{8}{25}.$

4.10.4   Finding of a Number by its Fraction

Problem. In a track-and-field sports event, with what speed should a long-distance runner run in order to cover a distance of 18 miles in $1\frac{1}{2}$ hours?

Solution. This problem may be solved by two methods. Both methods involve

<u>First method</u>

Since the long-distance runner should cover a distance of 18 miles in $1\frac{1}{2}$ hours (i.e., $\frac{3}{2}$ hours), then for $\frac{1}{2}$ hour he would cover a distance 3 times less than the required distance, that is $18 \div 3 = 6$ (miles); and for one hour, the long-distance runner would cover a distance 2 times more: $6 \cdot 2 = 12$ (miles):

$v = s \div t = 18 \div 1\frac{1}{2} = 18 \div \frac{3}{2} = (18 \div 3) \cdot 2 = \frac{18}{3} \cdot 2 = 18 \cdot \frac{2}{3} = 12$

(miles per hour);

Therefore, the speed necessary for the long-distance runner to cover a distance of 18 miles in $1\frac{1}{2}$ hours is 12 mph.

From the solution to this problem it is possible to deduce such a rule: *in order to find a number by the given value of its fraction, it is necessary to divide this value by the fraction.*

Second method

We designate the speed of the long-distance runner as *x mph*. Then for $1\frac{1}{2}$ hours, he would cover:

$$1\frac{1}{2} \cdot x = 18 \;\rightarrow\; \frac{3}{2} \cdot x = 18;$$

$$x = 18 \div \frac{3}{2} \rightarrow x = 18 \cdot \frac{2}{3} = 12 \text{ (mph)} - \text{speed of the long-distance}$$

runner.

Chapter Four: Questions to Test Your Understanding

1. What do fractional numbers (or mixed number fractions) express?

2. What is a common fraction?

3. What does the horizontal bar in a fraction indicate?

4. Define the following terms: (a) numerator; (b) denominator; (c) proper fraction; (d) improper fraction.

5. Give an example of each of the terms in question number 4.

6. Explain the relationship between the *quotient* obtained from the division of two natural numbers, and the *result* of a corresponding fraction.

7. What is the difference between the whole part and fractional part of a mixed number fraction?

8. How can the natural number 7 be written as a mixed number fraction having 3, 5, or 7 as the denominator?

9. How can two fractions having identical denominator be compared?

10. In terms of the values of their numerators and denominators, explain and give examples of how fractions compare with 1.

11. What are the rules that govern the addition and subtraction of fractions having identical denominators?

Chapter Four: Problems and Exercises

1. Write the fractions which are represented by the shaded and unshaded parts of each object below in Fig.4.5.

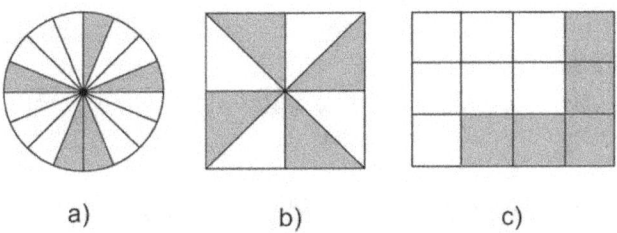

a)   b)   c)

Fig. 4.6. Fractional parts of objects

2. Draw a straight line having the length of 1 meter. Indicate on the straight line segments having the following lengths:

(a) $\frac{3}{10}$ of the given straight line;   (b) $\frac{4}{5}$ of the given straight line;

(c) $\frac{3}{4}$ of the given straight line;   (d) $\frac{7}{10}$ of the given straight line.

3. What fractional part of a day is: (a) 6 hours?   (b) 9 hours?   (c) 15 hours?

4. What fractional part of a dollar is each of the following coins: (a) 1 ¢ ; (b) 5 ¢ ; (c) 53 ¢ ; (d) 75 ¢ ?

5. What fractional part of a kilogram is: (a) 250 gram? (b) 700 gram? (c) 500 gram? (d) 850 gram?

6. Read the fractions below, show the numerator and denominator of each, and explain what they designate: (a) $\frac{2}{3}$;   (b) $\frac{3}{4}$;   (c) $\frac{4}{5}$;   (d) $\frac{5}{6}$;   (e) $\frac{6}{7}$;   (f) $\frac{3}{3}$.

7. The selling price of a new Toyota bus is $24,000. What is the selling price of a second hand taxicab, if its selling price is $\frac{1}{8}$ of the price of the Toyota bus?

8. Mariama paid $\frac{7}{12}$ of the money she had for her children's school fees. The money that remained with her after paying for the school fees was $720. Calculate how much money she had before paying the school fees?

9. Adokor had a yam which weighed 9 kg. She gave $\frac{2}{3}$ of it to Yarsah. What quantity of the yam did Adokor give to Yarsah?

10. A man walked $\frac{4}{7}$ of the distance between two villages. What is the distance between the villages if $\frac{4}{7}$ of it is equal to 28 km? Show a graphical illustration of your solution.

11. A group of 75 internally displaced people in Mount Barclay were given a food ration of 50-kg bag of rice as relief supply, which they equally divided among themselves. What fraction of the rice did each receive?

12. Write each of the following quotients as a fraction:
   (a) 7 m ÷ 4; (b) 3 T ÷ 5; (c) 5 h ÷ 8; (d) 5 kg ÷ 6;
   (e) 21 ÷ 11; (f) 19 ÷ 9.

12. Write each of the following quotients as a fraction:
   (a) 7 m ÷ 4; **(b)** 3 T ÷ 5; **(c)** 5 h ÷ 8; **(d)** 5 kg ÷ 6;
   (e) 21 ÷ 11; **(f)** 19 ÷ 9.

13. Doe was to celebrate his birthday with a party. So his parents bought a 6-kg cake for the party, amongst other things. Doe cut the cake into pieces so that it could be possible to equally share it among his friends; but it was not exactly known how many friends would be present. All he knew was that about 20 to 30 friends could attend. What is the least number of pieces into which Doe could cut the birthday cake so as to satisfy the condition of this problem?

14. Tanneh bought 17 oranges for 51 cents. What was the cost of one orange?

15. A motorist drove a car 3 miles for 45 minutes. What was the speed with which he drove?

16. A motorist is traveling between two cities X and Y. The distance between them is 450 miles, $\frac{5}{9}$ of which he has already covered. Calculate how many miles he has already covered, and show a graphical illustration of your answer.

17. Write the following in meters:

(a) (i) 3 dm; (ii) 17 dm; (iii) 23 dm; (iv) 9 dm;
(b) (i) 4 cm; (ii) 59 cm; (iii) 109 cm; (iv) 125 cm;
(c) (i) 17 mm; (ii) 32 mm; (iii) 303 mm; (iv) 1,005 mm.

18. Write the following in decimeters:
    (a) (i) 3 cm; (ii) 27 cm; (iii) 83 cm; (iv) 208 cm;
    (b) (i) 8 mm; (ii) 56 mm; (iii) 117 mm; (iv) 150 mm;
    (c) (i) 2 m 67 cm; (ii) 50 m 50 cm.

19. Write each of the following in hours as fraction:
    (a) (i) 3 min; (ii) 47 min; (iii) 75 min; (iv) 119 min;
    (b) (i) 3 sec; (ii) 150 sec; (iii) 500 sec; (iv) 5,050 sec.

20. The speed of a motorist is 72 kph (kilometers per hour). Express this speed in meters per second?

21. Write the following in hectares:
    (a) (i) 65 ares; (ii) 103 ares; (iii) 250 ares;
    (b) (i) 305 m^2; (ii) 1,500 m^2; 75 m^2.

22. Fill in the following Table 4.1 below, considering the relationship between a *hectare* an *are*.
            Table 4.1

Crop capacity of rice from 1 hectare	(a) 50 T	(b) 200 cnt	(c) 250 cnt	(d) 12,200 kg
Crop capacity of rice from 1 are				

23. Fill in the following Table 4.2 below:

                    Table 4.2

Kind of timber	(a)	(b)	(c)

	Ebony	Mahogany	Pine
Mass of 1 m³ of timber		1,500 kg	
Mass of 1 dm³ of timber	950 g		550 g

24. With which values of x the fraction $\frac{x}{15}$ is a proper fraction?

25. Write all the proper fractions having the denominator 10 and in which the numerators are whole numbers.

26. Convert or change each of the following improper fractions to mixed number fractions.

(a) $\frac{3}{2}$; (b) $\frac{15}{4}$; (c) $\frac{29}{7}$; (d) $\frac{85}{33}$; (e) $\frac{601}{50}$; (f) $\frac{201}{200}$.

27. Find the numbers by which it is possible to replace the letters x and y so that the equality could be true:

(a) $\frac{x}{25} = y + \frac{6}{25}$;   (b) $\frac{x}{18} = y + \frac{5}{18}$?

28. Find the number which, if divided by 23, would yield 6 as quotient and 11 as remainder.

29. Find the greatest two-digit number which if divided by 17 would yield 8 as remainder.

30. Draw a straight line, taking 4 cm as a unit length (i.e. 4 cm = 1), and mark the points which correspond to the following fractions:

(i) $\frac{1}{4}$; (ii) $\frac{2}{4}$; (iii) $\frac{3}{4}$; (iv) $\frac{4}{4}$; (v) $\frac{5}{4}$; (vi) $\frac{6}{4}$; (vii) $\frac{7}{4}$; (viii) $\frac{8}{4}$; (ix) $\frac{10}{4}$.

31. Arrange the following fractions in decreasing order, and tell which of them is the smallest and greatest:

(i) $\frac{8}{17}$; (ii) $\frac{18}{17}$; (iii) $\frac{13}{17}$; (iv) $\frac{20}{17}$; (v) $\frac{5}{17}$ (vi) $\frac{17}{17}$.

32. Supply the missing number that satisfies each equality:

189

(a) $\frac{13}{20}$ hour = _____ (minutes);

(b) 438 kg = _____ (tons);

(c) $\frac{41}{69}$ g = _____ (kg);

(d) _____ (centners) = $\frac{3}{10}$ ton;

(e) _____ meters = $\frac{7}{10}$ km;

(f) 28 cm = _____ m.

33. Compare each of the following pairs of fractions, indicating which is greater (or less) or equal, using the signs >, <, = :

(i) $\frac{13}{15}$ and $\frac{11}{15}$;

(ii) $\frac{9}{5}$ and $\frac{4}{5}$;

(iii) $\frac{9}{10}$ and $\frac{9}{13}$;

(iv) $\frac{7}{6}$ and $\frac{7}{5}$;

(v) $\frac{4}{5}$ and $\frac{6}{7}$.

34. Supply the values of the letter x which satisfies each inequality, using the signs < o r >:

(a) $\frac{11}{17} > \frac{11}{x}$;　　(b) $\frac{12}{25} < \frac{x}{25}$;　　(c) $\frac{8}{9} < \frac{8}{x}$;　　(d) $\frac{x}{15} > \frac{7}{15}$.

35. With what values of x the fraction $\frac{x}{15}$ would be less than the fraction $\frac{14}{15}$? Write at least a few of the fractions that are less than $\frac{14}{15}$.

36. Write a fraction that is three times less than each of the given ones:

(a) $\frac{2}{3}$ cm;　　(b) $\frac{3}{5}$ cm;　　(c) $\frac{1}{2}$ cm;　　(d) $\frac{6}{7}$ cm.

37. How does the value of a fraction change, if:
    (a) the numerator increases 4 times;

(b) the numerator decreases 10 times;

(c) the denominator increases 4 times;

(d) the denominator decreases 10 times?

38. How does the value of each of the following fraction change if its denominator is replaced by 3:

(a) $\frac{1}{6}$;   (b) $\frac{2}{9}$;   (c) $\frac{11}{15}$;   (d) $\frac{5}{9}$;   (e) $\frac{7}{21}$?

39. How many $\frac{2}{3}$ cm does 60 cm contain?

40. How many:

(a) $\frac{1}{4}$ meter^3 is contained in a volume of 12 m^3?

(b) $\frac{2}{5}$ liter is contained in 10 liters?

(c) $\frac{1}{5}$ hectare is contained in 15 hectares?

41. Calculate the following orally:

(a) $\frac{2}{7} + \frac{3}{7}$;   (b) $\frac{5}{8} - \frac{1}{8}$;

(c) $\frac{3}{11} + \frac{8}{11}$;   (d) $\frac{23}{30} - \frac{13}{30}$.

42. Find the value of each of the following expressions:

(a) $7 + 5\frac{2}{3}$;   (b) $17 - 12\frac{3}{4}$;

(c) $6\frac{4}{9} - 3\frac{5}{9}$;   (d) $9\frac{4}{7} - 5$.

43. Perform the additions by the most convenient method:

(a) $3\frac{4}{25} + 8\frac{2}{5} + 4\frac{3}{25} + 10\frac{3}{5}$;   (b) $5\frac{5}{8} + 7\frac{2}{5} + 1\frac{1}{5} + 3\frac{3}{8}$.

44. Give the answers in dollars, using fraction:

(a) 65 ¢ + 85 ¢ + 45 ¢;   (b) $1\frac{4}{5}$ dollars + 50 ¢ - 75 ¢.

45. Express the sum in kilograms:

(a) $\frac{4}{5}$ kg + 500 g + 2,060 g;   (b) $3\frac{1}{2}$ kg + 700 g.

46. Give the answers in minutes:

(a) $\frac{7}{12}$ hour + 45 minutes + 75 seconds;   (b) $2\frac{3}{4}$ hours + 30 minutes + 60 seconds.

47. Express the difference of $(35\frac{2}{5}$ km $- 25\frac{3}{5}$ km) in meters.

48. Solve the equations:

(a) $\frac{5}{8} + x = \frac{7}{8}$;

(b) $\frac{17}{20} - x = \frac{9}{20}$;

(c) $(\frac{6}{37} + x) - \frac{8}{37} = \frac{21}{37}$;

(d) $x - \frac{18}{49} = \frac{11}{49}$;

(e) $\frac{7}{19} + (3\frac{4}{19} - x) = \frac{9}{19}$;

(f) $\frac{19}{25} - (x + \frac{6}{25}) = \frac{4}{25}$.

49. Find the subtrahend, if the minuend is equal to $\frac{29}{31}$ and the difference is equal to $\frac{14}{31}$.

50. How does the sum of two numbers change if $4\frac{1}{6}$ is added to one and $2\frac{5}{6}$ is subtracted from the other?

51. How does the sum of two numbers change if $5\frac{3}{4}$ is added to each?

52. How does the difference of two numbers change if:

(a) The minuend is increased by $\frac{8}{9}$ and the subtrahend is increased by $\frac{4}{9}$?

(b)   The minuend is decreased by $\frac{4}{7}$ and the subtrahend is increased by $\frac{3}{7}$?

53. Yango has 45 balloons which are equal to $\frac{1}{3}$ of the number of balloons Akway has. Calculate how many balloons Akway has.

54. A steamer goes upstream on a river. The total speed of the steamer traveling upstream is $20\frac{4}{9}$ mph. The speed of flow of the river is $1\frac{2}{9}$ mph. Calculate the speed in mph of the steamer, if it is traveling downstream, assuming that downstream is in the direction against the flow of the river. Show graphical representation of your answer.

55. After Swen spent $\frac{2}{7}$ of his money and Tweh spent $\frac{4}{9}$ of his to buy some things in a supermarket, $55 still remained with each of them. How much more money Tweh had before than Swen?

56. If sugar cane contains $\frac{3}{4}$ part of sugar by mass, then what quantity of sugar cane is needed to produce 120 kg of sugar?

57. What quantity of rice should be milled in order to produce 6,000 kg of rice flour, if the mass of rice flour makes up $\frac{5}{6}$ of the mass of rice?

58. Which is greater: one-eighth of a quarter of $320, or one-half of half of $100.

59. How does the value of a fraction change, if its numerator is increased by the denominator?

60. How does the value of a proper fraction change, if both its numerator and denominator are increased by 1?

61. The denominator of a fraction is added to its numerator, and the numerator is added to its denominator. How does this affect the value of the original fraction?

62. How can 7 oranges be equally divided among 12 persons so that each orange is cut not more than 4 times?

63. A seller in a store sold $\frac{3}{5}$ of his goods before lunch time. After lunch time, he sold the remaining goods, which made up $15,000.00. Calculate what amount of goods in dollars he sold for that day.

Chapter Four: Exercises to Rate Your Ability

1. Solve the equations:
   (a) $\frac{7}{24} + y = \frac{23}{24}$;
   (b) $\frac{y}{67} - \frac{21}{67} = \frac{40}{67}$;
   (c) $(\frac{5}{51} + y) - \frac{12}{51} = \frac{19}{51}$;

2. A shop was supposed to sell 500 kg of mangoes in a day. Before lunch time it sold $\frac{1}{5}$ of the said quantity of mangoes. After lunch before closing time it sold additional $\frac{3}{5}$ of the mangoes. How many kilograms of mangoes the shop did not sell for that day?

3. Express the sum in ares: $\frac{4}{5}$ hectare + 35 ares + 50 m^2 .

4 With which values of y the fraction $\frac{y}{15}$ is less than the fraction $\frac{14}{15}$?

5 A technician repaired 170 machine parts he was working at. This is $\frac{10}{9}$ of his work quota. How many machine parts the technician might have repaired according to his work quota?

6 Perform the indicated operations:

(a) $\frac{5}{19} + \frac{1}{19} + \frac{7}{19}$;   (b) $7 - 3\frac{1}{8}$;

(c) $6 + \frac{5}{9}$;   (d) $\frac{7}{17} + \frac{9}{17} - \frac{12}{17}$.

7 Determine the whole numbers in the following fractions:

(a) $\frac{3}{11}$;   (b) $\frac{25}{14}$;

(c) $\frac{79}{13}$;   (d) $\frac{98}{17}$

8 With which values of a the fraction $\frac{a}{5}$ is a proper fraction?

9 Arrange the following fractions in the order of their increasing values: $\frac{3}{7}$; $\frac{8}{17}$; $\frac{16}{17}$; $\frac{4}{17}$; $\frac{15}{17}$; $\frac{9}{17}$; $\frac{5}{17}$; $\frac{6}{17}$.

10. Write the least improper fractions with the following denominators: (i) 2; (ii) 4; (iii) 6.

11. Write the least improper fractions with the following numerators: (i) 35; (ii) 75; (iii) 205.

12. Write the greatest proper fractions with the following denominators: (i) 15; (ii) 45; (iii) 125.

13. A secretary was instructed to type a 512-page document. For the first hour she typed $\frac{5}{16}$ of the said document; and for the second hour she typed $\frac{9}{16}$ of it. How many pages of the document did she type for the two hours?

14. A milk-man sold 48 liters of the milk he carried to the market to sell. The 48 liters is $\frac{4}{7}$ of his full container of milk. How many liters of milk were in his container of milk?

CHAPTER FIVE: DECIMAL FRACTIONS

5.1 The concept of Decimal Fractions

The denominator of a common fraction may be any natural number including 10 and also numbers such as 100, 1000, 10000, and so on. Fractions with such denominators are written to denote quantities already known to you, for example:

(a) 4 mm = $\frac{4}{10}$ cm;   (c) 1 dm = $\frac{1}{10}$ m;   (e) 15 cm = $\frac{15}{100}$ m ;

(b) 301 g = $\frac{301}{1000}$ kg ;   (d) 561 m = $\frac{561}{1000}$ km;   (f) 678 kg = $\frac{678}{1000}$ T.

It is easy to understand why the preceding expressions of equality are true, using your knowledge gained from the previous chapter. You already know that our system of counting uses decimal numeration or notation in which the unit of each rank (see Table 1.1, Chapter One) is 10 times larger than the unit of the preceding (smaller) rank. Conversely, the unit of each rank is 10 times smaller than the unit of the next (greater) rank. For natural numbers, it is known that the least rank is the *units rank*. But let us give it a second thought: "is there still a smaller rank or not?" Certainly, the answer can be "yes", if and when we are dealing with fractional numbers. In that case, the preceding rank according to our decimal system must be 10 times less than the unit of the next (greater) rank; that is, 10 times less than the *units rank* (or the *ones rank*). Consequently, a rank that is 10 times less than 1 or the units rank must be equal to 0.1 or $\frac{1}{10}$. You already know that this is one-tenth.   Therefore, the rank of tenths is conditionally written on the right of the rank of units. In order to show where the whole part of a fractional number (or mixed number fraction) ends, and where the fractional part begins, a point (commonly known as *decimal point*) is placed in front (i.e. on the left) of the fractional part.

For example, let us assume it is necessary to divide 6 pineapples among 10 persons. We can write or represent this division as $\frac{6}{10}$. It is obvious that not one of the ten persons will receive a whole pineapple; that is, it will be zero whole number and 6 tenths, or 0.6. Therefore, $\frac{6}{10}$ = 0.6. For the same reason, the following expressions are true:

(a) $15\frac{3}{10} = 15.3$; (c) $\frac{165}{100} = 1.65$; (e) $11\frac{2}{100} = 11.02$; (g) $6\frac{104}{1000} = 6.104$;

(b) $\frac{81}{100} = 0.81$; (d) $23\frac{3}{1000} = 23.003$; (f) $\frac{25}{100} = 0.25$;

(h) $987.3456 = 987\frac{3456}{10000}$.

The numbers *0.6, 15.3, 0.81, 1.65, 23.003, 11.02, 0.25, 6.104,* and *987.3456* are all *decimal fractions* (also called *decimals*). A *decimal fraction* (or *decimal*) *is a fraction that has a denominator of a power of ten, the power depending on (or deciding) the decimal place.* A decimal fraction is indicated by the *decimal point* to the left of the numerator, the denominator being omitted. A *decimal place is the position of a digit after the decimal point, each successive position to the right having a denominator of an increased power of ten.*

For example, in the decimal fraction 6.104, the digit 4 is in the third decimal place; and it has a denominator of $10^3 = 1000$. The digit 0 is in the second decimal place, having a denominator of $10^2 = 100$; the digit 1 is in the first decimal place and has a denominator of $10^1 = 10$:

$6.104 = 6 + \frac{104}{1000} = 6 + \frac{1}{10} + \frac{0}{100} + \frac{4}{1000} = 6 + 0.1 + 0 + 0.004 = 6 + 0.104 = 6.104$; or in words, *6.104 is equal to 6 whole number plus one tenth plus zero hundredth plus four thousandths (in short, 6 whole number one hundred and four thousandths).*

Thus, the term *decimal place* is also *used to refer to the number of digits to the right of the decimal point.* For instance, the number 987.3456 has four decimal places.

In view of the foregoing explanation and examples, the following statement is true:

decimal fractions are written on the basis of the same principle as natural numbers in the decimal (or base-ten) system which we are using: that is, each next rank from left to right is 10 times less than the preceding rank; or, conversely, each next rank from right to left is 10 times greater than the preceding rank.

Additional examples which further explain this concept are given below:

(a) $234.657 = 200 + 30 + 4 + 0.6 + 0.05 + 0.007 = 2 \cdot 100 + 3 \cdot 10 + 4 \cdot 1 + \dfrac{6}{10} + \dfrac{5}{100} + \dfrac{7}{1000};$

(b) $9{,}753.8642 = 9{,}000 + 700 + 50 + 3 + 0.8 + 0.06 + 0.004 + 0.0002 = 9 \cdot 1{,}000 + 7 \cdot 100 + 5 \cdot 10 + 3 \cdot 1 + \dfrac{8}{10} + \dfrac{6}{100} + \dfrac{4}{1000} + \dfrac{2}{10000};$

(c) $579 \text{ cm} = 5 \text{ m} + 70 \text{ cm} + 9 \text{ cm} = 5 \text{ m} + 7 \text{ dm} + 9 \text{ cm} =$

$= 5 \text{ m} + \dfrac{7}{10} \text{ m} + \dfrac{9}{100} \text{ m}.$

In order to write a decimal fraction in the form of a fractional number (or mixed number fraction), it is necessary:
- To write the non-zero number standing before the decimal point (i.e. on the left of the decimal point) as its whole number;
- To write the number standing after the decimal point (i.e. on the right of the decimal point) as the numerator; and then write the denominator as 1 followed by as many zeros as the number of decimal places. Zeros following directly after the decimal point are not included in the numerator.

For example:

(a) $234.657 = 234\frac{657}{1000}$; (c) $9{,}753.8642 = 9{,}753\frac{8642}{10000}$; (e) $17.354 = 17\frac{354}{1000}$;

(b) $0.46739 = \frac{46739}{100000}$; (d) $1.101 = 1\frac{101}{1000}$; (f) $0.0037 = \frac{37}{10000}$.

5.2 Properties of Decimal Fractions

The following properties are true for all decimal fractions (or decimals):

1. *The value of a decimal fraction does not change if any quantity of zeros is added from its right;*

    4.5 = 4.50 = 4.500 = 4.5000 = 4.50000, etc.

2. *The value of a decimal fraction increases by 10, 100, 1000, etc. times, if the decimal point is shifted (i.e. transferred or carried over) by one, two, three, etc. decimal places to the right;*

    (a) The number 7.6853 increases by 10 times, if the decimal point is transferred by one decimal place to the right, i.e. 76.853;

    (b) The number 7.6853 increases by 100 times, if the decimal point is transferred by two decimal places to the right, i.e. 768.53;

    (c) The number 7.6853 increases by 1000 times, if the decimal point is transferred by three decimal places to the right, i.e. 7685.3;

3. *The value of a decimal fraction decreases by 10, 100, 1000, etc. times, if the decimal point is shifted (i.e. transferred or carried over) by one, two, three, etc. decimal places to the left;*

    (d) The number 91384.2 decreases by 10 times, if the decimal point is transferred by one decimal place to the left, i.e. 9138.42;

    (e) The number 91384.2 decreases by 100 times, if the decimal point is transferred by two decimal places to the left t, i.e. 913.842 ;

(f) The number 91384.2 decreases by 1000 times, if the decimal point is transferred by three decimal places to the left, i.e. 91.3842;

5.3   Comparison of Decimal Fractions

The writing of decimal fractions follows the same pattern of writing natural numbers where the place value of each digit is important. Let us remember how natural numbers are compared, taking for example the numbers 8,975 and 8,985. It can be seen that the same concept is applicable in comparing the decimal fractions 8.975 and 8.985. The digits in the ranks of units and tenths of the two numbers are identical. If we look at their ranks of hundredths, the first number has 7 hundredths while the second has 8 hundredths. Since 8 hundredths is greater than 7 hundredths, the second number is greater than the first; that is, $8.985 > 8.975$, or $8.975 < 8.985$. Here two numbers with the same quantity of digits after the decimal point have been compared. But how can we compare similar numbers if the quantity of their digits after the decimal point is different? For instance, let us compare two numbers 53.82 and 53.81084. In the first number, we have only two digits after the decimal point, while in the second we have five digits after the decimal point. In both numbers the whole-number parts are equal. If we add some zeros from the right to a decimal fraction, then its value still remains unchanged. This means that adding three digits to the right of 53.82 would give us 53.82000. After adding the zeros, it has come to be that both numbers now have the same quantity of digits after the decimal point. Judging as in the previous example, we have $53.82000 > 53.81084$, or $53.82 > 53.81084$. Therefore, comparing the two numbers, it was not necessary by all means to have added the zeros; it was necessary simply to compare the digits in their hundredths place.

Consequently, in comparing decimal fractions, we use the same rule as in comparing natural numbers. In other words, decimal fractions are compared by means of the same place-value principle of their digit, starting from the greatest ranks. In general, the following applies:

- *Of two decimal fractions that one is greater (smaller) of which the whole number is greater (smaller);*
- *If their whole numbers are equal, then that decimal fraction is greater (smaller) of which the digit in the tenth place is greater (smaller);*
- *If their whole numbers and digits in the tenth places are equal, then that decimal fraction is greater (smaller) of which the digit in the hundredth place is greater (smaller); and so on.*

5.4   Rounding off of Decimal Fractions

Rounding off of decimal fractions occurs more frequently than rounding off natural numbers. It is not worth taking many decimal places in those cases when there is no certainty in their accuracy, or when there is no necessity in doing so. Accuracy of a measurement depends on the purpose for which it is made. For example, there are measurements which require adequate precision, and there are some which do not. The length of a textbook which is measured to be 24.53 cm may be rounded off to at least by 0.1 cm (that is, to the nearest tenth of centimeter).

In rounding off decimal fractions, we use the same rule as in rounding off natural numbers: if the first of the discarded or thrown off digit is 0, 1, 2, 3, or 4, then the last of the remaining digits is not changed; but if the first of the discarded digits is 5, 6, 7, 8, or 9, then the last of the remaining digits is increased by 1. The basic rule in rounding off natural numbers applies. See the following examples:

(a) 24.5359 ≈ 24.5 (rounded off to the nearest tenth);
(b) 24.5359 ≈ 24.54 (rounded off to the nearest hundredth);
(c) 24.5359 ≈ 24.536 (rounded off to the nearest thousandth);
(d) 24.5359 ≈ 25.0000 = 25 (rounded off to the nearest unit).

5.5   Addition and Subtraction of Decimal Fractions

The addition and subtraction of decimal fractions is basically the same as the addition and subtraction of common fractions. Let us look at the following problem, for example. The length of the sixth grade classroom is 17.75 meters, and that of the fifth grade is 19.55 meters. What is the total length in meters of both classrooms? The solution to such a problem requires the addition of the numbers 17.75 and 19.55:

$$17.75 + 19.55 = 17\frac{75}{100} + 19\frac{55}{100} = 36\frac{130}{100} = 37\frac{30}{100} \text{ (meters)}.$$

It is possible to solve the problem otherwise, by expressing the lengths in meters and centimeters:

17.75 m + 19.55 m = 17m 75cm + 19m 55cm = 36m + 130cm =37m 30cm = 37.30 m;

Or, working by addition in column, we can still have the same result:

$$\begin{array}{r} 17.75 \\ + 19.55 \\ \hline 37.30 \end{array}$$

Which of the methods of the solution to this problem do you consider more convenient, and why do you think so?

Let us look at another problem: By how many meters is the length of the fifth grade classroom (19.55 meters) greater than that of the sixth grade (17.75 meters)? In order to solve this problem, it is necessary to simply find the difference of the two numbers:

$$19.55 - 17.75 = 19\frac{55}{100} - 17\frac{75}{100} = 18\frac{155}{100} - 17\frac{75}{100} =$$
$$= 1\frac{80}{100} = 1.80 \text{ (meters)}.$$

Again, it is possible to solve this problem differently:

19.55 m − 17.75 m = 1955 cm − 1775 cm = 180 cm = 1.80 (meters);
or

$$\begin{array}{r} 19.55 \\ - 17.75 \\ \hline 1.80 \end{array}$$

From the preceding examples, it is obvious that the addition and subtraction of decimal fractions follows the same principle as the addition and subtraction of natural numbers. Unlike natural numbers, it is essential to pay attention so that the decimal points are correctly placed under each other in order to avoid making mistakes.

We had already discussed in Section 5.2 the case when decimal fractions have the same quantity of decimal places. However, there are other cases when it is required to add or subtract decimal fractions with different quantity of decimal places after the decimal point. In such instances it is accepted to add as many zeros as necessary after the decimal point to the decimal fraction with less quantity of decimal places before performing the required addition or subtraction operation. The following examples are intended to illustrate this concept:

1. 5.469075 + 209.832 =

$$\begin{array}{r} 5.469075 \\ + 209.832000^{*} \\ \hline 215.301075 \end{array}$$

000^{*}: Three zeros are added to complete the number of decimal places as in the first addend.

2. 425.1 − 89.50413 =

$$\begin{array}{r} 425.10000^{**} \\ - \quad 89.50413 \\ \hline 335.59587 \end{array}$$

203

0000^{**}: Four zeros are added to complete the number of decimal places as in the subtrahend.

3. $4.60531 - 2.918 =$

$$4.60531$$
$$- 2.91800$$
$$\overline{1.68731}$$

It should be noted that, however, that adding zeros to a fraction with less quantity of decimal places is not a matter of must. It is possible not to add or write the zeros at all. It is necessary only to mentally imagine them in those places where they are absent. See the next examples:

4. $23.32 - 21.94568 =$

$$23.32$$
$$- 21.94568$$
$$\overline{1.37432}$$

5. $35.53 + 4.26791 =$

$$35.53$$
$$+ 4.26791$$
$$\overline{39.79791}$$

6. $75.3 - 69.8579 =$

$$75.3$$
$$- 69.8579$$
$$\overline{5.4421}$$

It is easy to see in the preceding examples that the addition of decimal fractions are carried out according to ranks: tens are added to tens, units to units, tenths to tenths, hundredths to hundredths, and so forth. The same is true during the subtraction of decimal fractions. Therefore, in the addition (subtraction) of decimal fractions, it is important to write one rank under the other so that identical ranks of addends (minuend and subtrahend) and the decimal points are correctly positioned one under the other. Additions and subtractions are performed according to ranks starting with the least rank from the right and paying attention to the decimal point. In the case of different quantity of decimal places, one should be reminded that the necessary number of zeros are added to equalize the number of decimal places; or one ought to simply mentally imagine the necessary quantity of zeros in those places where they are absent. Hence, an addition operation of any number of addends can be performed, for example:

7. $14.051 + 235.172635 + 1.04 + 0.9081 =$

$$
\begin{array}{r}
14.051 \\
235.172635 \\
1.04 \\
+ \quad 0.9081 \\
\hline
251.171735
\end{array}
$$

5.6 Multiplication of Decimal Fractions

The multiplication of decimal fractions is not different from the multiplication of common fractions. As in ordinary decimal operations, it is important to be watchful of the quantity of decimal places in the factors involved so as to take stock of them in the product. The following examples help to illustrate the concept of multiplying decimal fractions:

1. Problem. The length of a bedroom is 15.25 m and its width is 6.65 m. What is the area of the given bedroom?

Solution. In order to determine the area (*A*) of the bedroom, it is necessary to multiply the length (15.25 m) and width (6.65 m):

$$A = l \cdot w = 15.25 \cdot 6.65 = 101.4125 \ (\text{m}^2)$$

2. Problem. What is the volume of the bedroom in the problem above, if it is 2.73 m high?

Solution. In order to find the volume (V) of the bedroom, we should multiply the length (l), width (w), and height (h):

$$V = l \cdot w \cdot h = 15.25 \cdot 6.65 \cdot 2.73 = 276.856125 \ (m^3)$$

What does it mean to be watchful of the quantity of decimal places in the factors during the multiplication of decimal fractions? As the foregoing examples have illustrated, there are basic rules that apply: *to multiply two or more decimal fractions, it is necessary in the product to count from the right and separate by a decimal point as many decimal places as the total number of decimal places in all of the factors.*

3. <u>Problem.</u> What is the volume of a rectangular solid whose length, width, and height are respectively 0.15 cm, 0.07 cm, and 0.4 cm?

<u>Solution.</u> $V = l \cdot w \cdot h = 0.15 \cdot 0.07 \cdot 0.4 = 0.0042 \ (cm^3)$. Here in this example, we first perform the multiplication $15 \cdot 7 \cdot 4$ to get 420. We notice that there are five decimal places in the three factors (0.15, 0.07, and 0.4); so we count five decimal places in the product (420) starting from the right. This gives us 0.00420. But the last zero on the right of the 2 is not significant; therefore it can be dropped or discarded without affecting the value of the product.

4. <u>Problem.</u> Perform the indicated multiplication: 0.21 x 0.09 x 0.12.

<u>Solution.</u> We first multiplied 0.21 by 0.09 to get 0.0189; and then we multiplied 0.0189 by 0.12 to get the product 0.002268. This is the same as multiplying $21 \cdot 9 \cdot 12$ to get 2268. We notice that the total number of decimal places in the three factors (0.21, 0.09, and 0.12.) is six; so we counted six decimal places in the product. See below the indicated operations:

$$
\begin{array}{rr}
0.21 & \quad 0.0189 \\
\text{x } \underline{0.09} & \text{x } \underline{\quad 0.12} \\
0.0189 & 378 \\
& + \underline{\quad 189} \\
& \underline{0.002268}
\end{array}
$$

5. Problem. Perform the indicated multiplication: 0.354 x 0.27.

Solution. We multiplied 0.354 by 0.27 to get the product 0.09558. This is the same as multiplying 354 · 27 to get 9558. We notice that the total number of decimal places in the two factors (0.354 and 0.27.) is five; so we counted five decimal places in the product. See below the indicated operation.

$$
\begin{array}{r}
0.354 \\
\times \ \underline{0.27} \\
2478 \\
+ \ \underline{708} \\
\underline{0.09558}
\end{array}
$$

6. Problem. Perform the indicated multiplication: 0.563 x 82.

Solution. We multiplied the decimal fraction 0.563 by the natural number 82 to get the product 46.166. This is the same as multiplying 563 · 82 to get 46166. We notice that the number of decimal places in the decimal fraction 0.563 is three. Since a natural number does not contain decimal places, the number of decimal places in 82 is zero. So we counted three decimal places in the product. See below the indicated operation.

$$
\begin{array}{r}
0.563 \\
\times \ \underline{82} \\
1126 \\
+ \ \underline{4504} \\
\underline{46.166}
\end{array}
$$

In the multiplication of decimal fractions, all the properties (commutative, associative, and distributive properties) of multiplication you had already studied in Chapter Three are true and applicable.

5.7 Multiplication and Division of Decimal Fractions by 10, 100, and 1000 Times

You already know that adding zeros from the right to a decimal fraction does not affect its value. See the following examples:

1. $0.75 = 0.750 = 0.7500 = 0.75000 = \dfrac{75}{100} = \dfrac{750}{1000} = \dfrac{7500}{10000} = \dfrac{75000}{100000}$
= and so on;

2. $0.0042 = 0.00420 = 0.004200 = \dfrac{42}{10000} = \dfrac{420}{100000} = \dfrac{4200}{1000000} =$ and so on.

On the contrary, each zero added from the right to a natural number increases the value of the given natural number by ten times. Thus, in multiplying a natural number (39 for instance) by 10, a zero does appear in the product from the right: $39 \times 10 = 390$. The digit 9 which initially stood in the units place now stands in the tens place; and the digit 3 which initially stood in the tens place now stands in the hundreds place. Consequently, because of a zero added from its right as a result of multiplying it by ten, the preceding ranks of a natural number would appear to have shifted by one place to the left. If two zeros were added from the right (as in the case of multiplying by 100), then the preceding ranks of a natural number would appear to have shifted by two places to the left. And so on.

Let us suppose it is required to increase the decimal fraction 7.3 by ten times. It is obvious from the previous example that we should have to shift or move the ranks by one position to the left. For this reason, it is sufficient only to shift the decimal point by one digit to the right. You can be convinced in this by the following examples:

1. (a) 7.346 m = 73.46 dm = 734.6 cm = 7346 mm;
   (b) 7346 mm = 734.6 cm = 73.46 dm = 7.346 m.

Multiplication (or Increase) of Decimal Fractions by 10, 100, 1000, Etc.:

2. (a) $7.346 \cdot 10 = 73.46$ (the decimal point is shifted one decimal place to the right);

(b) $7.346 \cdot 100 = 734.6$ (the decimal point is shifted two decimal places to the right);

(c) $7.346 \cdot 1{,}000 = 7346$ (the decimal point is shifted three decimal places to the right);

(d) $7.346 \cdot 10{,}000 = 73460$ (the decimal point is shifted four decimal places to the right).

Division (or Decrease) of Decimal Fractions by 10, 100, 1000, Etc.:

3. (a) $7346 \div 10 = 734.6$ (the decimal point is shifted one decimal place to the left);

(b) $7346 \div 100 = 73.46$ (the decimal point is shifted two decimal places to the left);

(c) $7346 \div 1{,}000 = 7.346$ (the decimal point is shifted three decimal places to the left);

(d) $7346 \div 10{,}000 = 0.7346$ (the decimal point is shifted four decimal places to the left).

In accordance with the trend of reasoning in the above examples, we can make the following conclusion:

- *in order to increase (or multiply) a decimal fraction by 10, 100, 1000 (and so on), it is sufficient only to shift the decimal point from left to right by one, two, three (and so on) decimal places, respectively.*

- *in order to decrease (or divide) a decimal fraction by 10, 100, 1000 (and so on), it is sufficient only to shift the decimal point from right to left by one, two, three (and so on) decimal places, respectively*

A decimal fraction can also be seen as a quotient from the division of natural numbers. This is exactly the case when one natural number is finitely divided by another natural number. See the examples below.

(a) $13 \div 8 = 1.625$;

$$
\begin{array}{r}
1.625 \\
8\overline{)13.000} \\
-\underline{8} \\
50 \\
-\underline{48} \\
20 \\
-\underline{16} \\
40 \\
-\underline{40} \\
0
\end{array}
$$

(b) $15 \div 4 = 3.75$;

$$
\begin{array}{r}
3.75 \\
4\overline{)15.00} \\
-\underline{12} \\
30 \\
-\underline{28} \\
20 \\
-\underline{20} \\
0
\end{array}
$$

(c) $9 \div 5 = 1.8$.

$$
\begin{array}{r}
1.8 \\
5\overline{)9.0} \\
-\underline{5} \\
40 \\
-\underline{40} \\
0
\end{array}
$$

In each of the examples given above, the division operation was performed exactly, i.e. without a remainder. However, exact division without a remainder is not always possible no matter how long we try to continue the operation. In such cases, since the division continues infinitely and/or repeatedly, it is accepted to approximate or round off the value of the quotient to a defined nearest decimal rank (to the nearest tenths, hundredths, and so on, for instance). In this way, the quotient is said to be an ***approximate quotient***. Some examples of *approximate quotients* are given below:

(a) $10 \div 3 = 3.333\ldots$;

$$
\begin{array}{r}
3.333 \\
3\overline{)10.000} \\
-\underline{9} \\
10 \\
-\underline{9} \\
10 \\
-\underline{9} \\
10 \\
-\underline{9} \\
1
\end{array}
$$

(b) $5 \div 9 = 0.555\ldots$;

$$
\begin{array}{r}
0.555 \\
9\overline{)5.000} \\
-\underline{45} \\
50 \\
-\underline{45} \\
50 \\
-\underline{45} \\
5
\end{array}
$$

(c) $8 \div 6 = 1.33\ldots$;

$$
\begin{array}{r}
1.33 \\
6\overline{)8.00} \\
-\underline{6} \\
20 \\
-\underline{18} \\
20 \\
-\underline{18} \\
2
\end{array}
$$

(d) $22 \div 7 = 3.142857\ldots$;

$$
\begin{array}{r}
3.142857 \\
7\overline{)22.000000} \\
-\underline{21} \\
10 \\
-\underline{7} \\
30 \\
-\underline{28} \\
20 \\
-\underline{14} \\
60 \\
-\underline{56} \\
40 \\
-\underline{35}
\end{array}
$$

$$\begin{array}{r} 50 \\ -\underline{49} \\ 1 \end{array}$$

The three dots at the end of each quotient in the examples above signify that the given digit or train of digits repeats infinitely. This is obviously the case with the repeating train of digits *.142857* in example (d) above. In such cases, division is discontinued and limited to some first digits of the quotient, having preliminarily rounded it off to a defined decimal rank.

Consequently, the quotient from the division of two natural numbers may be *finite* or *infinite*. That is why decimal fractions may be *finite* or *infinite*. *Infinite decimal fractions* are also referred to as **repeating decimal fractions**. As we have seen, examples of **infinite (or repeating**) decimal fractions are 3.333..., 0.555..., 1.33..., 3.142857, and many others.

5.8   Division by Decimal Fractions

Let us suppose that it is required for us to divide 47.965 by 9.05. Relatively, this may not be so simple to do. So we try to reduce it to a case of dividing by a natural number with which we are already acquainted. For this reason, we should remember a basic property of division: *if a dividend and a divisor are multiplied by the same number, then the quotient always remains unchanged.* Therefore, we will multiply the divisor and the dividend by a certain number such that the divisor becomes a natural (whole) number which will simply our problem. Such a number would be 100, since 9.05 x 100 = 905; and 47.965 x 100 = 4796.5. It is easy to see that the new numbers (4796.5 and 905) were gotten simply by transferring or shifting the decimal point two decimal places to the right of the original numbers. Thus the division by a decimal fraction is reduced to a division by a natural number: *47.965 ÷ 9.05 = 4796.5 ÷ 905 = 5.3*:

$$905 \overline{)4796.5}^{\,5.3}$$
$$-\underline{4525}$$
$$2715$$
$$\underline{-2715}$$
$$0$$

Consequently, the basic rule that applies is that: *in order to divide by a decimal fraction, it is necessary to shift the decimal point to the right in both the divisor and the dividend as many decimal places as are in the divisor, and after that perform the division by a natural number.* Some examples are given below:

(a) $1.24 \div 0.005 = 1240 \div 5 = 248;$

$$5 \overline{)1240}^{\,248}$$
$$-\underline{10}$$
$$24$$
$$-\underline{20}$$
$$40$$
$$-\underline{40}$$
$$0$$

(b)

$2679.264 \div 672 = 3.987;$

$26.79264 \div 6.72 =$

$$672 \overline{)2679.264}^{\,3.987}$$
$$-\underline{2016}$$
$$6632$$
$$-\underline{6048}$$
$$5846$$
$$-\underline{5376}$$
$$4704$$
$$-\underline{4704}$$
$$0$$

$$(c) 23.46 \div 0.46 = 2346 \div 46 = 51;$$

$$\begin{array}{r} 51 \\ 46\overline{)2346} \\ -\underline{230} \\ 46 \\ -\underline{46} \\ 0 \end{array}$$

5.9   Arithmetic Mean

Mulbah sold for five days in the shop of his father Mr. Kpadeh who was away attending other pressing matters. When Mr. Kpadeh returned, he was curious and wanted to know at once how much money was realized from the sales by his son during his absence. So he asked Mulbah what was the average amount of sales he made per pay for the five days he was away. Mulbah answered his father that for the five days he realized the following sales: $60 + $75 + $90 + $150 + $110. Obviously, this was not the exact answer that Mr. Kpadeh had expected. He wanted to know the average sales per day, then he could determine right away the total amount for the five days Mulbah had worked. For this, it was necessary for Mulbah to find the total sales for the five days ($60 + $75 + $90 + $150 + $110 = $485) and divide by 5. That is, *$485 ÷ 5 = $97.* So the average sales per day which Mr. Kpadeh wanted to know from his son was $97.

Here the number *485* designates the sum of certain number or quantity of addends; the *5* is the number or quantity of addends; and *97* is *the average or arithmetic mean* of the 5 addends whose sum is 485.

Therefore, the rule for finding an arithmetic mean is this: *in order to find the arithmetic mean of a given quantity of numbers, it is necessary to divide the sum of these numbers by their quantity.* In other words, *arithmetic mean is the average value of a set of integers, numbers, or items, expressed as their sum divided by their quantity.* Arithmetic mean is widely used in practical life situations. It is especially applicable not only in mathematics but also in physics,

214

chemistry, biology, and in a variety of scientific and technological processes.

Chapter Five: Questions to Test Your Understanding

1. (a) What is a *decimal fraction*?
   (b) In your own words, briefly explain the difference between a *fractional number* and a *decimal fraction*.

2. How the writings of decimal fractions and natural numbers do compare?

3. How a decimal fraction can be written in the form of a fractional number (or mixed number fraction)?

4. What rules do you know for comparing decimal fractions?

5. What rules do you know for rounding off decimal fractions?

6. Give an example of adding and another example of subtracting decimal fractions.

7. Give an example of multiplying two decimal fractions.

8. Give some examples of multiplying a decimal fraction by 10, 100, and 1000.

9. Give some examples of dividing a decimal fraction by 10, 100, and 1000.

10. Give an example of a decimal fraction as a quotient from the division of two natural numbers.

11. What adjectives are used to qualify two kinds of decimal fractions that result from the division of two natural numbers?

12. Give one example of finite and another example of infinite (repeating) decimal fractions.

13. What is an *arithmetic mean* or *average*? Give an example.

14. In your opinion, why the concept of *arithmetic mean* is necessary, and where can it be applied?

Chapter Five: Problems and Exercises

1. Write each number in the form of a sum of ranked units, considering the place value of every digit:
    (a) 127,951   (c) 237,865
    (b) 348,693   (d) 461,094

2. Expand each fraction into addends. Write the numerators in the form of ranked units, expressing the denominator as a power of ten; and then write the resulting sum as a decimal fraction.
    (a) (i) $\frac{61}{10}$; (ii) $\frac{415}{100}$; (iii) $\frac{602}{100}$;
    (b) (i) $\frac{1235}{1000}$; (ii) $\frac{60947}{10000}$; (iii) $\frac{950041}{100000}$.

3. Perform the indicated addition of decimal fractions:
    (a) $56 + 0.56 + 0.034 + 0.0072$;
    (b) $91 + 0.08 + 0.009$;
    (c) $143 + 0.06 + 0.003 + 0.0001$.

4. Write each of the following in dollars and then express as a fraction or fractional number (mixed number fraction).
    (a) (i) 1,289 cents; (ii) 242 cents; (iii) 87 cents; (iv) 2 cents;
    (b) (i) 78 cents; (ii) 201 cents; (iii) 615 cents; (iv) 95 cents.

5. Express the sum in meters and then write as a fractional number:
    (a) 35 m + 2 dm + 6 cm + 5 mm;
    (b) 23 m + 1 dm + 1 cm + 0 mm;
    (c) 0 m + 0 dm + 95 cm + 8 mm;
    (d) 11 m + 7 dm + 7 cm + 7 mm.

6. Write the sum as a decimal and then express as a fractional number:
 (a) 15 whole number + 9 tenths;
 (b) 79 whole number + 7 tenths + 23 hundredths;
 (c) 55 whole number + 1 tenth + 85 hundredths + 3 thousandths.

7. Write the given quantities in tons, using decimal fractions:
 (a) 7,564 kg;
 (b) 867 kg;
 (c) 98 kg;
 (d) 5 kg;
 (e) 0.4 kg

8. Write in square meters:
 (a) 92 dm^2;
 (b) 35 dm^2;
 (c) 62 cm^2;
 (d) 85 cm^2;
 (e) 6 dm^2 + 3 cm^2.

9. Draw a number line, and mark on it a segment AB with length 0.9 dm from the beginning (i.e. from its zero point). Write the length of the segment in millimeters, centimeters, meters, and kilometers.

10. Compare each pair of decimal fractions, using the symbols =, > and <:
 (a) 5.1 and 4.5;
 (b) 6.73 and 3.987;
 (c) 2.356 and 2.653;
 (d) 32.53 and 32.530;
 (e) 7.650 and 7.659;

11. Tell whether each of the following expressions is true or false. Write the sign "$\neq$" if the values are not equal:
 (a) 35.90 > 35.9;
 (b) 0.7154 > 0.75;

(c) $5.550 = 5.55$;
(d) $0.0006 > 0.06$;
(e) $8.439 < 8.394$.

12. Find at least one value of y which satisfies each of the following inequalities:
   (a) $65 > y > 64$;
   (b) $11 < y < 11.003$;
   (c) $5.9 < y < 6$;
   (d) $0.8 < y < 0.9$;
   (e) $0.123 < y < 0.124$.

13. Read each of the following decimal fractions and tell which of them is the greatest and which of them is the least: (i) 6.7; (ii) 0.67; (iii) 0.067; (iv) 0.0067; (v) 0.00067.

14. Arrange the decimal fractions in order of their decreasing values: 0.36; 0.149; 0.0502; 0.143523; 0.712; 0.513; 0.145; 0.61.

15. Arrange the decimal fractions in problem 14 in order of their increasing values.

16. Round off the given decimals.
   (a) To the nearest units: (i) 0.89; (ii) 0.35; (iii) 0.009; (iv) 17.012; (v) 2.65; (vi) 0.719; (vii) 84.5;
   (b) To the nearest tenths: (i) 4.33; (ii) 35.011; (iii) 1.019; (iv) 0.514; (v) 9.05;
   (c) To the nearest hundredths: (i) 2.6780; (ii) 5.3055; (iii) 0.0897; (iv) 21.7456; (v) 1.0934;
   (d)   To   the   nearest
      thousandths: (i) 3.8024; (ii) 5.6789; (iii) 0.01038; (iv) 4.98725; (v) 0.2009.

17. Read the approximate equalities and determine to which place-value rank each number has been rounded off.
   (a) $73.725 \approx 73.73$;
   (b) $73.725 \approx 73.7$;

(c) $73.725 \approx 74$;
(d) $17.6457 \approx 17.65$;
(e) $17.645 \approx 18$.

18. Find the sum or difference.
    (a) $46.5 - 29.651$;
    (b) $248.136 - 159.759$;
    (c) $23.297 + 4.908$;
    (d) $1.235 + 0.876$

19. Write the answers in dollars.
    (a) $4 ¢ + 49 ¢ + 20 ¢$;
    (b) $203 ¢ + 73 ¢ + 97 ¢$.

20. Write the answers in meters per minute (m/min):
    (a) 30 km/hr $- 15,000$ m/hr $+ 0.35$ km/min;
    (b) 12 km/hr $- 0.18$ km/min;
    (c) 7.5 km/hr $+ 2.4$ km/min $+ 80$ m/sec.

21. Write the sum in kilograms:
    (a) 3.4 T $+ 17$ Cnt $+ 5$ kg;
    (b) 0.807 T $+ 0.923$ Cnt $+ 61$ kg.

22. Write the sum in *ares (a)*:
    (a) 5.8 ha $+ 8.5$ ha $+ 29$ m^2;
    (b) 0.65 ha $+ 91$ a $+75$ m^2.

23. Perform the indicated operations:
    (a) $(78.537 - 11.618) - (18.086 + 2.573)$;
    (b) $7.1 - 2.51 + (0.152 - 0.095)$;
    (c) $(13.145 + 2.9) - 6.05$.

24. Find y, if:
    (a) $y + 31.94 = 55.09$;
    (b) $15 - y = 6.42$;
    (c) $(23.1 - 4.3) - y = 12.6$;
    (d) $0.0239 + y = 18.4$.

25. How does the sum of two numbers change if one is increased by 9.3 and the second is decreased by 5.4?

26. The difference of two numbers is equal to 0.849. What would their new difference be if the minuend is increased by 6.4 and the subtrahend also increased by 1.87?

27. One side of of a triangle is equal to 25.7 cm, the second side is 4.1 cm shorter than the first, and the third is 8.2 cm shorter than the second. What is the perimeter of this triangle?

28. Find the length of the third side of a triangle, if the first side is equal to 24.1 cm and the second side is 5.5 cm greater than the first. What can be said about the value of the perimeter of this triangle?

29. A median EG is drawn in an isosceles triangle DEF with base DF. (A median is a vertical line connecting the top of a triangle and the middle of the opposite side). Find the length of the median EG, if the perimeter of the triangle DEF is 70 cm, and that of triangle DEG is 60 cm.

30. Think of any number. Multiply it by 7, and add a number $k = 42$ to the product. Divide the sum by 7, and subtract from the quotient the number you had initially thought of. The result will be 6. Explain why. Make up a formula or expression for calculating and checking the result.

31. Three workers were to receive a pay of $560.80 for a certain work which they did. One of them worked 10 hours; the second worked 4 hours, and the third worked 6 hours. What pay could each worker receive for the portion of work done by him?

32. For three days a motorist traveled a distance of 589 miles with a constant speed. The first day he drove for 5 hours, the second day – 6 hours, and the third day – 8 hours. How many miles did the motorist travel on the third day more than on the first?

33. The speed of a motor torpedo-boat traveling downstream (in the direction of river current) on a river is 15.8 mph, and the speed of river flow is 1.9 mph. Find the boat's speed upstream (against the direction of river flow). What would be the speed of the boat, if the speed of flow of the river were zero?

34. The own speed of a motor boat on a river is 19.7 mph. The speed of flow of the river is 2.4 mph. Find the speed of the boat when it is traveling upstream (against the direction of river flow) and downstream (in the direction of river flow).

35. Calculate the following by the most convenient method you can think of.
    (a) $8 \cdot 0.7 \cdot 0.25$;
    (b) $16 \cdot 0.15 \cdot 0.125$;
    (c) $2.7 \cdot 11.8 + 4.1 \cdot 2.7$;
    (d) $0.12 \cdot 17.3 - 15.4 \cdot 0.12$;
    (e) $60 \cdot 0.14 \cdot 0.3$.

36. Calculate orally:
    (a) $1.5 \cdot 0.3$;   (d) $3.5 \cdot 20$;
    (b) $0.15 \cdot 4$;   (e) $8 \cdot 0.08$;
    (c) $0.2 \cdot 0.5$;   (f) $3 \cdot 0.9$.

37. Perform the indicated operations and write the answers in square meters (m^2):
    (a) 7.3 ares $\cdot$ 5.2;
    (b) 46 $dm^2 \cdot$ 11.8;
    (c) 105 $dm^2 \cdot$ 4.7.

38. One side of a square is equal to 16.5 cm. Find its perimeter and area.

39. From which number must 5.4 be subtracted in order to get a number that is 2.7 times greater than 8.6?

40. The length of a rectangle is equal to 12.3 m, and its width is 0.7 of the length. Find the perimeter and area of such a rectangle.

41. By how much the area of a rectangle with sides 14.2 cm and 11.4 cm is greater (or smaller) than the area of a square having a side 12.6 cm?

42. Solve the following problems:

(a) A field having an area of 90 hectares was planted with cassava, rice, and eddoes. The cassava occupied 0.35 of all the area; the rice occupied 25.5 hectares more than cassava. How many hectares did the eddoes occupy?

(b) 0.45 of money available was used to buy 8 kg of bananas costing \$2.25 per kg; the rest of the money was used to buy oranges. How much money was used to buy the oranges?

43. Calculate and compare the areas and perimeters of rectangles MNOP, if:
(a) MN = 2.3 cm, NO = 2.7 cm;
(b) MN = 2.5 cm, NO = 2.5 cm;
(c) MN = 1.8 cm, NO = 3.2 cm;
(d) MN = 6 cm, NO = 1.5 cm;
(e) MN = 3 cm, NO = 3 cm;
(f) MN = 5 cm, NO = 1.8 cm.

44. Calculate the product $m \cdot n$ and the sum $m + n$, if:
(a) m = 4.4, n = 1.5;
(b) m = 2.3, n = 5.6;
(c) m = 1.2, n = 2.1.

45. It is necessary for the Monrovia Consolidated School System (MCSS) to purchase 1,950 tons of corn meal from three stores. The quantity of corn meal which each store sells and the cost of transportation in dollars are given in Table 5.1 below. Make up a purchasing plan of the corn meal so that the cost of its transportation would be the least.

Table 5.1

Stores	1	2	3
Quantity of corn meal available for purchase, tons	700	1,900	650
Transportation cost per ton of corn meal	$5.25	$7.75	$3.50

46. Orally calculate the following:
    (a) $9.85643 \cdot 1000$;
    (b) $0.5 \cdot 10,000$;
    (c) $0.0134 \cdot 100$;
    (d) $0.67 \cdot 0.01$;
    (e) $0.04 \cdot 0.001$.

47. Write in meters:
    (a) 0.56 dm;
    (b) 53.8 km;
    (c) 0.1234 km;
    (d) 75.86 dm.

48. How many kilograms are in each of the following quantities?
    (a) 0.31 T;
    (b) 4.78 Cnt;
    (c) 345.9 g?

49. Answer the following questions:
    (a) What is the equivalent of 470 m^2 in *Ares*?
    (b) What is the equivalent of 9.6 hectares in Ares?
    (c) By how many times is the number 71.64 greater than the number 0.7164?
    (d) By how many times is it necessary to increase the number 2.43 in order to get 243?
    (e) By how many times is it necessary to decrease the number 5.67 in order to 0.0567?

50. Solve the equations:
    (a) $(x + 5) - 8 = 65$;
    (b) $45 - 12x = 17$;
    (c) $20 + 3x = 51$.

51. A bus traveled 384 miles in 5 hours. What is the speed of the bus if its motion was uniform?

52. The perimeter of a parcel of land having the form of a square is equal to 96 m. Find its area.

53. The area of a rectangle is 540 m^2, and one of its sides is 27 m. Determine its perimeter.

54. Find the following:
    (a) 24 oranges is what fraction of 96 oranges?
    (b) 17 kg is what fraction of 51 kg?
    (c) The product of two numbers is 115. One of them is 11.5. Find the other number.

55. A sheet of lead with length 22 cm and width 12 cm has a mass of 135 kg. Determine the mass of 10 cm^2 of this sheet of lead.

56. The cost of subscription to a fitness club is $84.60 for one year. What is the subscription cost for 3 months?

57. Find the length of a container if its height is 3 m, width 5 m, and volume 90 m^3.

58. A roll of cloth 36.2 m in length should be cut into 2 parts such that one part is 6.2 m greater than the other. What would be the length of each part of the roll of cloth?

59. Two reservoirs initially contained equal quantity of fuel oil. After 860 liters of fuel oil were taken from the first, the quantity of fuel oil in the second became 5 times as great as the fuel oil quantity which

remained in the first. What was the quantity of fuel oil initially contained in each reservoir?

60. From Robertsport City (west of Liberia) and Harper City (east of Liberia) two buses departed at the same time to travel in opposite directions towards each other. (The distance between the two cities is assumed to be 360 miles). The speed of one of the buses was 25 mph, and that of the other was 35 mph. With which speed did the two buses approach each other? What distance was between them after 2 hours? After how many hours did the buses meet upon departure from their respective starting points?

61. An automatic machine can manufacture 479 machine parts in one hour. In what time can it manufacture 750 machine parts under the same working condition?

62. Calculate the quotient and round off to the nearest hundredths:
    (a) $0.41 \div 33$;
    (b) $0.75 \div 183$;
    (c) $15 \div 156$;
    (d) $23 \div 415$.

63. Round off the following numbers to the nearest units:
    (a) 123.456;
    (b) 4.93;
    (c) 87.56;
    (d) 19.198.

64. Round off the numbers in problem number 63 (a) to (d) to the nearest tenths.

65. Solve the equations:
    (a) $11.1 \div x = 25.4$;
    (b) $3 \div x = 0.345$;
    (c) $8.5 - 5x = 3.4$;
    (d) $2.65 \div x = 0.356$.

66. Kojo has $25 which is 0.25 of the money Kofi has. How much money does Kofi have?

67. After a worker spent 0.65 of his salary, he still had $75 left. What is the salary of the worker?

68. The area of a rectangular parcel of land is 250 square miles. If one of its sides is 25 miles, what is its perimeter?

69. If the length and width of the rectangular parcel of land in problem 68 above were 25 miles and 10 miles respectively, by how many times its area would increase if its width were increased by 5 miles?

70. The map of a stadium in a rectangular form has dimensions of 15 mm x 18 mm. Determine its actual dimensions and area, if the scale on the map is 1:9250.

71. Find the average speed of a motorist, if he drove to town with the speed of 54 mph and retuned with the speed of 45 mph. The distance to town is 81 miles.

72. The weights of 4 bags of bananas were determined to be 11.5 kg, 13 kg, 9,9 kg, and 15.6 kg. What is the average weight of the 4 bags?

73. The speed of a motor boat going along with the stream current of a river is 15.7 mph. Its speed when going against the stream is 11.5 mph. Calculate the own speed of the boat and the speed of flow of the river current.

74. The *arithmetic mean* of two numbers (x and 26) is 18.4. Find the value of x.

75. Geegbae earned the following marks in 7 exams: Mathematics – 89; Geography – 98; History – 76; English – 79; French – 83; Literature – 91, and Biology – 85. What is Geegbae's average for the 7 exams.

76. From a bus station, two buses departed in opposite directions at the same time. The speed of one was 58.2 km per hour, and that of the other was 69.3 km per hour. What was the distance between them after 3 hours?

77. The distance between two motor cyclists traveling towards each other is 421 miles. The speed of one of them is 38.5 mph, and that of the other is 45.7 mph. Within what time will the two motor cyclist meet?

78. The number of tourists in one bus was 3 times more than the number of tourists in a second bus. When 8 tourists left the first bus to board the second, the number of tourists in both buses became equal. How many tourists initially were in each bus?

79. The area of a triangle is 185 m². Find the base of the triangle, if its height is 37 m.

80. The number of oranges in one bag was 5 times more than those in a second bag. After 24 oranges were taken from the first and added to the second bag, the quantity of oranges in both bags became equal. How many oranges originally were in each bag?

Answers to Problems and Exercises

Chapter One

1. (a) 64,195. (b) 2,014. (c) 2,655,230. (d) 1,357. **2.** 343. **3.** 625. **4.** 5. **5.** 10. **6.** $x = 8$. **7.** 3. **8.** 24. **9.** 27. **10.** Only (b) 1,502 and (d) 1,284. **11.** Only (b) 5,600 and (c) 9,532. **12.** Only (c) 2,652. **13.** All of them. **14.** Only (d) 9,051. **15.** Only (b) 1,978 and (c) 1,239. **16.** All of them. **17.** 13. **18.** The least four-digit number is 1000; and the greatest two-digit number is 99. Their difference is 901. **19.** 23. **20.** 35 lb of fruits more than vegetables. **21.** The three preceding numbers are $x-3$, $x-2$, $x-1$; and the three succeeding numbers are $x+1$, $x+2$, $x+3$. **22.** The sum of any three consecutive natural numbers is not divisible by 6 either when the first of them is even or the second of them is odd. **23.** (b) One billion three hundred nine millions four hundred fifty-two thousands eight hundred seventy-nine. (d) One billion and fourteen. **24.** (a) 543,793,207,829. (c) 45,000,000,333. **25.** (d) Five hundred thousands, five units. (e) Fifty thousands, five thousands, five hundreds, five tens. **26.** 555; 559; 599;955; 959; 995; 595; 999. **27.** 973 – the greatest; 379 – the least; **28.** 135; 153; 315; 351; 513; 531. Their sum is 1,998. **29.** Pupil Y (5 minutes) – the first; Pupil X (7 minutes) – the second; Pupil Z (9 minutes) – the last. **30.** (c) $a = 2,209$; (d) $d = 4,729$. **31.** 20 hours – the time required for the bicyclist to cover the distance traveled by the train in 4 hours. **33.** 300,605,079. **34.** 3,418. **35.** (a) Rounded off to the nearest hundreds; (b) rounded off to the nearest tens; (c) rounded off to the nearest tens of thousands; (d) rounded off to the nearest hundreds of thousands. **36.** (a) $864,097 \approx 864,100$ – rounded off to the nearest tens; (d) $864,097 \approx 860,000$ – rounded off to the nearest tens of thousands. **37.** (a) 24,600; (b) 15,400; (c) 74,000; (d) 597,000. **38.** 60,000. **39.** 6 kg. **40.** (a) (i) 25,000 km; (ii) 4,000 km; (iii) 4,000 km; (iv) 10,000 km; (b) (i) \$54; (ii) \$9; (iii) \$25; (iv) \$3. **43.** Their sum is 8 cm; and their difference is 2 cm. **44.** PD ≈ 4.75 cm. **45.** OPTR = 12 cm. **46.** 1,205 miles – the distance from City Y to City Z through City X. **47.** A point which is 1 inch to the left of point M, and another point which is 1 inch to the right of point N. **50.** 36 cm. **51.** 38 ft. **52.** (a) <; (b) >; (c) >; (d) >. **53.** (a) <; (b) >; (c) =. **56.** The greatest: 555,441; the least: 144,555.

Chapter Two

1. Yes, the sum of $x + y$ can be equal to x provided that y is equal to zero. **2.** (a) The result is equal to 5,000; (b) The result is equal to 2,300. **3.** (a) (40 +60) + 30 > 100 − 30; (b) (150 − 90) + 95 < 60 + 100; (c) (175 − 50) + 40 < 200 + 40; (d) (30 + 40) + 75 < 100 + 75. **4.** $(x − y) > 145,000$. **5.** MNOPST = 15 inches. **6.** (a) \$6.70; (b) 2,551 km. **7.** The sum does not change. **8.** 2,030 miles. **9.** (a) x = 37; (b) y = 58. **10.** The difference of (567 − 312) is decreased by 100; (b) The difference of (567 − 312) is increased by 100. **11.** (a) x = 504; (b) y = 40; (c) x = 6; (d) y = 18. **12.** The difference between the roots of the given equations is 6. **13.** The cost of a ticket is 70 cents. **14.** The price of the graduation present was $225. **15.** $534 - (312 - 199) > 845 - (200 + 245)$. **16.** The two parts of the 517 cents would be 47 cents and 47 dimes. **17.** For five seconds the parachutist descended 330 feet. **18.** The false cigars were in the fifth pack, since five cigars were taken from there. **19.** The textbook: $54; the copybook: $5; and the school bag: $76. **20.** (a) 29,994; (b) 1; (c) 89. **21.** (a) 232; (b) 550; (c) 20,200. **22.** (a) inches; (b) kg; (c) dm; (d) cents. **24.** (a) $x = 35$ – second addend; (b) $x = 396$ – first addend; (c) $x = 569$ – sum; (d) $x = 114$ – subtrahend; (e) $x = 180$ – minuend; (f) $x = 475$ – difference. **25.** (i) $199; (ii) $176. **26.** (a) Adding the same number to the minuend and subtrahend in a subtraction operation does not change the difference; (b) The same explanation as in (a); (c) Subtraction of a sum from a number; (d) Rounding-off method; **27.** 6,597,550. **28.** 64,290. **29.** (c) 4567; (d) 1234. **30.** 381 hectares; **36.** (a) 90^0; (b) 180^0; (c) 60^0; (d) 90^0. **37.** (a) ½; (b) 1/6; (c) ¼; (d) 1/3. **38.** (a) 19 hours; (b) 18 hours 30 minutes; (c) 18 hours 15 minutes; (d) 18 hours 10 minutes. **43.** 3 inches; **44.** The perimeter of the triangle is 3 units greater than the perimeter of the rectangle. **47.** $P_\Delta = 12$ cm; $L_{1st\ side} = 3$ cm; $L_{2nd\ side} = 4$ cm; $L_{3rd\ side} = 5$ cm; **48.** $P_\Delta = (x + y + z)$ in. **49.** 15 in. $< y < 25$ in. **51.** Both cars would arrive at their destinations at the same time. **52.** (a) 52 centners; (b) 33 centners; (c) 9 centners. **53.** (a) 59 km 102 m; (b) 4 kg 970 g; (c) 7 hours 7 minutes. **55.** 42,743. **56.** − 8,280. **58**. 96. **59.** (a) 31 km 675 m –sum; 10 km 375 m – difference; (b) 8 tons 100 kg – sum; 2 tons 700 kg – difference; (c) $1,415.54 – sum; $395.82 – difference. **60.** 65. **61.** (a) (i) $x < 3$; (ii) $x \le 3$; (iii) $x > 3$; (b) (i) $x > 10$; (ii) $x \ge 10$; (iii) $x < 10$; (c) (i) $x < 17$; (ii) $x \le 17$; (iii) $x > 17$; (d) (i) $x > 5$; (ii) $x \ge 5$; (iii) $x < 5$. **62.** (a) $x = 165$; (b) $x = 358$; (c) $a = 3,957$; (d) $y = 272$; (e) $x = 7$; (f) $x = 10$; (g) $y = 400$. 63. (a) $x = 63$; (b) $x = 200$. **64.** $x = 5$. **65.** (a) $x = d \div 3$ (oranges); if $d = 42$ cents, then $x = 14$ oranges; (b) $x = 16 \cdot t$ (mimeographs); if $t = 22$ seconds, then 352 mimeographs. **66.** 4 kg. **68.** (a) 3,600 km; (b) 900 km; (c) 1,200 km; (d) 1,500 km. **69.** (i) 7 inches – length of St. Paul River on map; (ii) 5 inches – length of Sanquin River on map. **70.** (a) 7 kg 420 g, or 7.24 kg; reverse problem: a bag of rice weighing 50 kg is heavier than a bag of farina by 7 kg 420 g. What is the weight of the bag of farina? (b) 31 plums; reverse problem: Tenesee picked 96 plums and gave 31 of them to Tanneh. How many plums still remained with him? **71.** 92. **72.** 51. **73**. 4. **74.** 12. **75.** 8. **76.** 97,682. **77.** 4,476. **79.** 106 ducks; 74 geese; 25 chicks. **80.** $x < 7$. **81.** (a) 3; (b) 4. **82.** (a) $x = 190$; (b) $x = 31$; (c) $x = 40$. **83.** (a) rice – 120 kg (given); corn – 30 kg; cassava – 50 kg; eddoes – 40 kg; total – 240 kg; (b) $(160 + 6x)$ kg of seed rice. **84.** Since Juagbeh was more far from the Telecoms Building, it meant she was closer to the Finance Ministry Building because she walked slowly. It is obvious that Doe walked faster than Juagbeh.

Chapter Three

1. (a) $534 \cdot 4 = 2{,}136$; (b) 138; (c) $4n$; (d) nk. **2.** (a) 25; (b) $3n$; (c) 9. **3.** (a) 1 and 11; (b) 5 and 0; (c) 3 and 5; (d) 4 and 5. **4.** (a) $64n$ miles; (b) 256 miles; (c) 320 miles; (d) 448 miles. **5.** (b) $x = 5$ or $x = 3$; (c) $x = 4$; (d) $x = 3$. **6.** (a) (i) not correct; (iii) correct; (b) (ii) correct; (iv) not correct. **7.** (a) \$2; (b) 15 yards; (c) $3\frac{1}{3}$ times. **8.** (a) (i) 1,800; (ii) 1,500; (iii) 18,000; (b) (iii) 39; (iv) 180; (v) 2. **9.** (a) (i) 2,000 kg; (ii) 300 kg; (iii) 0.45 kg; (b) (i) 45 tons; (ii) 5 tons; (c) (i) $(n \div 100)$ centners; (ii) k centners; (iii) $10a$ centners. **10.** (a) 10,000; (b) 3. **11.** 15 hours. **12.** 810 bottles of soft drinks; $54x$ – number of soft drink bottles brought the next day. **13.** (a) $30a$; (b) $84kn$; (c) $40x^2$. **14.** (ii) 24 yards^2; (iv) 28 yards^2; (vi) 90 yards^2. **15.** 4 rectangles; the rectangle with dimensions of 4 yards and 6 yards. **17.** (a) 486,000; (d) 8,850,000,000. **19.** 45 miles. 20. 572 miles. **21.** (a) 1,311 seats; (b) \$2,164.80; (c) \$955.20. **22.** (a) 180; (c) $60 + 15a$; (e) $3m + 3n$. **23.** 354 miles. **24.** (a) 3,717,287; (b) 5,618,394; (c) 5,313,98. **25.** (a) (i) 12,682; (b) (i) 16,122. **26.** 600 miles. **27.** 209 miles – distance traveled by the motor boat on both the lake and downstream the river for seven hours; 169 miles - distance traveled by the motor boat on both the lake and upstream the river for seven hours. **28.** (b) 1,876 (quotient) + 344 (remainder). **29.** (a) $x = 290$; (b) $m = 1$; (c) $a =$ any number. (d) $n = 2$. **30.** (a) 5; (b) 25; (c) $6n$; (e) a. **31.** (a) 105; (c)20; (e) 720; (f) 2. **32.** (a) 4; (b) 750; (c) 5; (d) 53. **33.** 447. **34.** (a) 9; (b) 21; (c) 6; (d) 6. **35.** The father's age is 30; the son's is 5. **36.** The first number is 72; the second is 9. **37.** (a) 50,682; (b) 901,325; (c) 50,557. **38.** 17. **40.** (a) 9 mph; (b) 9 textbooks. **41.** (a) $x = 105$; (b) $x = 580$. **42.** (a) $4{,}032x$; (b) $253y$. **43.** 20 mph. **44.** 1. **46.** 47 m^2. **47.** The product would increase by 5 times. **48.** 8,145. **49.** 108. **50.** The quotient would increase 30 times. **53.** The diameter of the circle is 60 mm; the radius is 30 mm. **55.** 4.1 cm – the greatest distance between points D and B. 60. Perimeter = 46 cm; Area = 120 cm^2. **61.** 56 cm^2. **62.** 361 cm^2. **63.** 17,600 cm^2. **64.** Four of such rectangles are possible. The rectangle with length of 8 cm and width of 1 cm has the least area. **65.** 10 m^2. **66.** 80 poles. **67.** 3 cm^2. **68.** 90 cm^2 – initial area; 54 cm – perimeter of the new rectangle. **69.** 144 cm^2. **70.** 1,008 cm^2. **71.** 5 cm. **72.** 120 cm – the sum of the lengths of all the edges of the cube; 600 cm^2 – the area of the total surface of the cube. **73.** 60.5 cm^2. **74.** 729 cm^3. **75.** 1,134 kg. **76.** 7 kg.

Chapter Four

1. (a) $\frac{5}{16}$ (shaded); $\frac{11}{16}$ (unshaded); (b) $\frac{4}{8} = \frac{1}{2}$ (shaded); $\frac{4}{8} = \frac{1}{2}$ (unshaded); (c) $\frac{5}{12}$ (shaded); $\frac{7}{12}$ (unshaded). **2.** (a) 30 cm; (b) 80 cm; (c) 75 cm; (d) 70 cm. **3.** (a) $\frac{1}{4}$; (b) $\frac{3}{8}$; (c) $\frac{5}{8}$. **4.** (a) $\frac{1}{100}$; (b) $\frac{1}{20}$; (c) $\frac{53}{100}$; (d) $\frac{75}{100} = \frac{3}{4}$; **5.** (a) $\frac{250}{1000} = \frac{1}{4}$; (b) $\frac{700}{1000} = \frac{7}{10}$; (c) $\frac{500}{1000} = \frac{1}{2}$; (d) $\frac{850}{1000} = \frac{17}{20}$. **7.** \$3,000.00. **8.** \$1,728.00. **9.** 6 kg. **10.** 49 km. **11.** $\frac{2}{3}$ (kg). **12.** (a) $1\frac{1}{4}$ (m); (b) $\frac{3}{5}$ (T); (c) $\frac{5}{8}$ hour; (d) $\frac{5}{6}$ (kg); (e) $1\frac{10}{11}$; (f) $2\frac{1}{9}$. **13.** Since it was expected that either 20 or 30 friends could attend the birthday party, Doe needed to cut his birthday cake into $20 \cdot 30 = 600$ pieces. Each piece would weigh 6000 (g) $\div$ 600 (pieces) = 10 g per piece. If 20 friends came to the party, then each could receive 600 (pieces) $\div$ 20 = 30 (pieces); if 30 friends came to the party, then each could receive 600 (pieces) $\div$ 30 = 20 (pieces). **14.** 3 ¢. **15.** $\frac{1}{15}$ (mile per minute). **16.** 250 miles. **17.** (a) (i) $\frac{3}{10}$ m; (iii) $2\frac{3}{10}$ m; (b) (ii) $\frac{59}{100}$ m; (iv) $1\frac{1}{4}$ m; (c) (i) $\frac{17}{1000}$ m; (iv) $1\frac{1}{200}$ m. **18.** (a) (ii) $2\frac{7}{10}$ dm; (iii) $8\frac{3}{10}$ dm; (b) (i) $\frac{2}{25}$ dm; (iv) $1\frac{1}{2}$ dm; (c) (i) $26\frac{7}{10}$ dm; (ii) 505 dm. **19.** (a) (i) $\frac{1}{20}$ hour; (iii) $1\frac{1}{4}$ hours; (b) (ii) $\frac{1}{24}$ hour; (iv) $1\frac{29}{72}$ hours. **20.** 20 meters per second. **21.** (a) (i) $\frac{13}{20}$ hectare; (iii) $2\frac{1}{2}$ hectares; (b) (i) $\frac{61}{2000}$ hectare; (iii) $\frac{3}{400}$ hectare. **22.** (a) 500 kg = 5 cnt; (b) 200 kg; (c) 250 kg; (d) 122 kg. **23.** (a) 950 kg; (b) 1.5 kg; (c) 550 kg. **24.** $0 < x < 15$. **25.** $\frac{1}{10}, \frac{2}{10}, \frac{3}{10}, \ldots, \frac{9}{10}$. **26.** (a) $1\frac{1}{2}$; (b) $1\frac{1}{14}$; (c) $4\frac{1}{7}$; (d) $2\frac{19}{33}$; (e) $12\frac{1}{50}$; (f) $1\frac{1}{20}$. **27.** (a) If $y = 0$, then $x = 6$; (b) If $y = 0$, then $x = 5$. **28.** 149. **29.** 93. **31.** $\frac{20}{17}, \frac{18}{17}, \frac{17}{17}, \frac{13}{17}, \frac{8}{17}, \frac{5}{17}$. **32.** (a) 39 minutes; (b) $\frac{219}{500}$ T; (c) $\frac{41}{69000}$ kg; (d) 3 Cnt; (e) 700 m; (f) $\frac{7}{25}$ m = 0.28 m.

33. $\frac{13}{15} > \frac{11}{15}$; $\frac{9}{5} > \frac{4}{5}$; $\frac{9}{10} > \frac{9}{13}$; $\frac{7}{5} > \frac{7}{6}$. **34.** (a) $x > 17$; (b) $x > 12$; (c) $x < 9$; (d) $x > 7$. **35.** $0 < x < 14$. **36.** (a) $\frac{2}{9}$ cm; (b) $\frac{1}{5}$ cm; (c) $\frac{1}{6}$ cm; (d) $\frac{6}{21}$ cm. **37.** (a) Its value increases by 4 times; (b) Its value decreases by 10 times; (c) Its value decreases by 4 times; (d) Its value increases by 10 times. **38.** (a) Its value increases by 2 times; (b) Its value increases by 3 times; (c) Its value increases by 5 times; (d) Its value increases by 3 times; (e) Its value increases by 7 times. **39.** 90. **40.** (a) 48; (b) 25; (c) 75. **42.** (a) $12\frac{2}{3}$; (b) $4\frac{1}{4}$; (c) $2\frac{8}{9}$; (d) $4\frac{4}{7}$. **43.** (a) $26\frac{7}{25}$; (b) $17\frac{3}{5}$. **44.** (a) $\$1\frac{95}{100} = \$1.95$; (b) $\$1\frac{55}{100} = \$1.55$. **45.** (a) $3\frac{9}{25}$ kg or 3.36 kg; (b) $4\frac{1}{5}$ kg. **46.** (a) $81\frac{1}{4}$ min; (b) 196 min. **47.** 9,800 meters. **48.** (a) $x = \frac{1}{4}$; (c) $x = \frac{23}{37}$; (e) $x = 3\frac{2}{19}$; (f) $x = \frac{9}{25}$. **49.** $\frac{15}{31}$. **50.** The sum is increased by $1\frac{1}{3}$.

51. The sum is increased by $11\frac{1}{2}$. **52.** (a) The difference will increase by $\frac{4}{9}$; (b) The difference will decrease by 1. **53.** 135. **54.** $19\frac{2}{9}$ mph. **55.** $22.00. **56.** 160 kg; **57.** 7,200 kg. **58.** one-half of half of $100.00. **59.** The fraction increases by 1. **60.** The value of the proper fraction increases. **61.** The new fraction would be equal to 1. **62.** *Three oranges* could be divided into four equal parts; and the remaining *four oranges* could be divided into three equal parts. **63.** $37,000.

Chapter Five

1. (a) $127,951 = 100,000 + 20,000 + 7,000 + 900 + 50 + 1$; (d) $461,094 = 400,000 + 60,000 + 1,000 + 90 + 4$. **2.** (a) (i) $\frac{61}{10} = \frac{60}{10} + \frac{1}{10} = 6 + \frac{1}{10} = 6.1$;

(b) (i) $\frac{1235}{1000} = \frac{1000}{1000} + \frac{200}{1000} + \frac{30}{1000} + \frac{5}{1000} = 1 + \frac{2}{10} + \frac{3}{100} + \frac{5}{1000} = 1.235$;

(iii) $\frac{950041}{100000} = \frac{900000}{100000} + \frac{50000}{100000} + \frac{40}{100000} + \frac{1}{100000} = 9 + \frac{5}{10} + \frac{4}{10000} +$

$\dfrac{1}{100000}$ = 9.50041. **3.** (a) 56.6012; (c) 143.0631. **4.** (a) (i) $12.89 = $12\dfrac{89}{100}$;

(b) (iv) $0.95 = $\dfrac{95}{100}$. **5.** (a) 35.265 m = $35\dfrac{265}{1000}$ m; (c) 0.958 m = $\dfrac{958}{1000}$ m; (d)

11.777 m = $11\dfrac{777}{1000}$ m. **6.** (b) 79.93 = $79\dfrac{93}{100}$; (c) 55.953 = $55\dfrac{953}{1000}$. **7.** (a)

7.564 T; (c) 0.098 T; (e) 0.0004 T. **8.** (b) 0.35 m^2; (d) 0.0085 m^2; (e) 0.0603 m^2. **10.** (c) 2.356 < 2.653. **11.** (a) 35.90 =35.9; (c) True. **12.** (a) y = 64.5; (c) y = 5.95. **13.** (i) six whole number and seven tenths; (iii) zero whole number, six hundredths, and seven thousandths. **16.** (**a**) (i) 1; (iii) 0; (v) 3; (vii) 5; (**b**) (ii) 35; (iv) 0.5; (v) 9.1; (**c**) (i) 2.68; (iii) 0.09; (v) 1.09; (**d**) (i) 3.802; (iii) 0.01; (v) 0.201. **17.** (a) seventy-three whole number seven hundred twenty-five thousandths is approximately equal to seventy-three whole number seventy-three hundredths (*rounded off to the nearest hundredths*); (c) seventy-three whole number seven hundred twenty-five thousandths is approximately equal to seventy-four whole number (*rounded off to the nearest units*); (e) seventeen whole number six hundred forty-five thousandths is approximately equal to eighteen whole number (*rounded off to the nearest units*).

18. (a) 16.849; (c) 28.205. **19.** (a) $0.73; (b) $3.73. **20.** (a) 600 m/min; (c) 7,325 m/min. **21.** (a) 5,105 kg. **22.** (b) 156.75 *ares*. **23.** (a) 46.26; (c) 9.995. **24.** (a) y = 23.15; (d) y = 18.3761. **25.** The sum is increased by 3.9. **26.** Their new difference would be increased by 4.53. **27.** 60.7 cm. **28.** 5.5 cm < x < 53.7 cm, where x is the length of the third side. 59.2 cm < P < 107.4 , where P is the perimeter of the triangle. **29.** 25 cm. **31.** $280.40 – the pay for the first worker; $112.16 – the pay for the second worker; $168.24 – the pay for the third worker. **32.** 93 miles. **33.** 13.9 mph – the speed of boat in still water; 12 mph – speed of boat traveling upstream. **34.** 22.1 mph – the speed of boat traveling downstream; 17.3 mph – the speed of boat going upstream. **35.** (a) 1.4; (b) 0.3; (d) 0.228; (e) 25.2. **37.** (a) 3,796 m^2; (c) 4.935 m^2. **38.** 66 cm – perimeter; 272.25 cm^2 – area. **39.** 28.62. **40.** 41.82 m – perimeter; 105.903 m^2 – area. **41.** 3.12 cm^2 – as much the area of the rectangle is greater than the area of the square. **42.** (a) 1.5 ha; (b) $22. **44.** (a) 6.6 - product; 5.9 – sum; (c) 2.52 – product; 3.3 – sum. **47.** (a) 0.056 m; (d) 7.5 m. **48.** (b) 478 kg; (c) 0.3459 kg. **49.** (a) 4.7 *ares;* (b) 960 *ares;* (c) 100 times; (d) 100 times. **50.** (a) x = 68; (c) x = $10\dfrac{1}{3}$. **51.** 76.8 mph. 52. 576 m^2.

53. 94 m. **54.** (a) $\dfrac{1}{4}$; (c) 10. **55.** 5.1 kg. **56.** $21.15. **57.** 6 m. **58.** 15 m – length of the shorter part; 21.2 m – length of the longer part. **59.** 1,075 liters. **60.** (a) 60 mph – the speed with which the two buses approached each other; (b) 240 miles – distance between buses after 2 hours; (c) 6 hours – the time within which the buses met after departure from their respective starting points. **61.**

1.57 (hour) – time required to manufacture 750 machine parts. **62.** (a) 0.01; (b) 0; (d) 0.06. **63.** (c) 88; (d) 19. **64.** (a) 123.5; (b) 4.9. **65.** (b) x = 8.69565; (d) x = 7.4438. **66.** $100. **67.** $214.29. **68.** 70 miles. **69.** 1.5 times. **70.** 138.75 m x 166.5 m – the actual dimensions of the stadium; 23, 101.875 m^2 – the actual area of the stadium. **71.** 49.09 mph – average speed of motorist. **72.** 12.5 kg. **73.** 13.6 mph – speed of motor boat; 4.2 mph – speed of flow of river. **74.** x = 10.8. **75.** 85.86 **76.** 382.5 km. **77.** 5 hours. **78.** 8 – number of tourists in second bus; 24 - number of tourists in first bus. **79.** 10 m.

APPENDIX

I. Tables of Weights and Measures

The Metric System

A. Linear Measure (or Units of Length)

1 mm = 0.03937 in
10 mm = 1 cm = 0.3937 in
10 cm = 1 dm
10 dm = 100 cm = 1 m = 39.37 in =
3.2808 ft

1,000 m = 1 km = 0.621 mile =
3,280.8 ft
1 dm = 100 mm
1 m = 1,000 mm
1 km = 100,000 cm

B. Square Measure (or Units of Area)

$1 mm^2 = 0.00155 in^2$
$100 mm^2 = 1 cm^2 = 0.15499 in^2$
$100 cm^2 = 1 dm^2$
$100 dm^2 = 1 m^2 = 1,549.9 in^2 =$
$1.196 yd^2$
$1,000,000 m^2 = 1 km^2$

$100 m^2 = 1 are (a) = 119.6 yd^2$
100 a = 1 hectare (ha)
100 square hectometers $(hm^2) = 100$
$ha = 1 km^2 = 0.386 mile^2 = 247.1$
acres
$1 acre = 4,046.86 m^2 = 4,840 yd^2$

C. Volume Measure (or Units of Volume)

$1,000 mm^3 = 1 cm^3 = 0.06102 in^3$
$1,000 cm^3 = 1 dm^3 (1 liter) = 61.023$
$in^3 = 0.0353 ft^3$
10 liters = 1 decaliter
100 liters = 1 hectoliter

$1,000 dm^3 = 1 m^3 = 35.314 ft^3 =$
$1.308 yd^3$
$1,000,000 cm^3 = 1 m^3$
$1,000,000,000 m^3 = 1 km^3$

D. Weight/Mass Measure (Units of Weight or Mass)

10 decigrams (dg) = 1 gram (g) =
15.432 grains = 0.035274 ounce
(avdp);
1,000 mg = 1 g
1,000 g = 1 kg

10 hectograms (hg) = 1 kg = 2.2046
pounds
100 kg = 1 centner (cnt)
10 cnt = 1,000 kg = 1 ton (T)
10 quintals = 1 metric ton = 2204.6
pounds

E. Land Measure

100 centiares = 1 are (a) = 119.6 yd^2
100 a = 1 hectare (ha) = 2.471 acres

100 ha = 1 km^2 = 0.386 mile^2 =
0.386 mile^2 = 247.1 acres

The British (or Imperial) System

A. Linear Measure (or Units of Length)

1 mil = 0.001 in = 0.0254 mm
1 in = 1,000 mils = 2.54 cm
12 in = 1 ft = 0.3048 m
3 ft = 1 yd = 0.9144 m
$5\frac{1}{2}$ yd = $16\frac{1}{2}$ ft = 1 rod (or pole) =
5.029 m

40 rods (or poles) = 1 furlong =
201.168 m
8 furlongs = 1760 yd = 5280 ft = 1
mile = 1.6093 km

B. Square Measure (or Units of Area)

1 in^2 = 6.452 cm^2
144 in^2 = 1 ft^2 = 929.03 cm^2
9 ft^2 = 1 yd^2 = 0.8361 m^2
$30\frac{1}{4}$ yd^2 = 1 rod^2 (or pole^2) = 25.292 m^2

160 rod^2 = 4840 yd^2 = 43,560 ft^2 = 1
acre = 0.4047 ha
640 acres = 1 mile^2 = 259.00 ha =
2.590 km^2

C. Cubic Measure

1 in^3 = 16.387 cm^3
1728 in^3 = 1 ft^3 = 0.0283 m^3

27 ft^3 = 1 yd^3 = 0.7646 m^3

II. Useful Equivalents

1 mile = 1760 yards = 1.609
kilometers (km)
1 km = 0.621 mile
1 inch = 2.54 centimeters (cm)
1 cm = 0.3937 inch
1 meter (m) = 39.37 inches = 3.28
feet (ft) = 1.09 yard
1 yard = 0.9144 meter (m)
1 kg = 2.204 pounds (lb)

1 pound = 0.453 kg= 16 ounce (oz)
1 oz = 28.35 grams (g)
1 hour = 60 minutes = 3,600 seconds
100 m^2 (square meters) = 1 are (a)
100 ares = 1 hectare
1 hectare = 10,000 m^2 = 2.471 acres
1 m^2 = 100 dm^2 = 10,000 cm^2
1 km^2 = 1,000,000 m^2
I gallon (gal) = 3.79 liters

III. List of Abbreviations and Symbols

a = are(s)
ans = answer
avdp =avoirdupois
kg = kilogram
mm = millimeter
m = meter
m^2 = square meter(s)
m^3 = cubic meter(s)
mi = mile(s)
mi^2 = square mile(s)
sq= square
g = gram
 lb = pound
oz = ounce
cm = centimeter
hr = hour
min = minute
sec = second
mph = miles per hour
kph = kilometer per hour

yd = yard(s)
yd^2 = square yard(s)
yd^3 = cubic yard(s)
km = kilometer
$\in$ = belongs to (is a member of)
$\angle$ = angle
$\perp$ = perpendicular
$\parallel$ = parallel
$\triangle$ = triangle
$\leq$ = less than or equal to
$\geq$ = greater than or equal to
$\approx$ = almost equal to
$>$ = greater than
$<$ = less than
^0C = degree (s) Celsius
e.g. = for example
esp. = especially

ha = hectare(s)
T = ton(s)
cnt = centner(s)
dm = decimeter
etc. = et cetera
ft = foot (feet)
ft^2 = square foot (feet)
ft^3 = cubic foot (feet)
i.e. = that is
in = inch(es)
in^2 = square inch(es)
in^3 = cubic inch(es)
pt= point(s)
dg = decigram
hm = hectometer
hm^2 = square hectometer
hm^3 = cubic hectometer
hg = hectogram
... = and so on

IV. The Greek Alphabet

Capital letter	Small letter	Name of symbol	Associated meaning or value
A	α	Alpha	The first
B	β	Beta	The second
Γ	γ	Gamma	
Δ	δ	Delta	
E	ε	Epsilon	
Z	ζ	Zeta	
H	η	Eta	
Θ	θ	Theta	

I	ι	Iota	The smallest part
K	κ	Kappa	
Λ	λ	Lamda	
M	μ	Mu	
N	ν	Nu	
Ξ	ξ	Xi	
O	o	Omicron	
Π	π	Pi	≈ 3.14
P	ρ	Rho	Specific resistance
Σ	σ	Sigma	Summation
T	τ	Tau (Tao)	
Y	υ	Upsilon	
Φ	φ	Phi	
X	χ	Chi (Khi)	
Ψ	ψ	Psi	
Ω	ω	Omega	The last

V. A Table of Common Polygons and Polyhedrons and Their Formulae

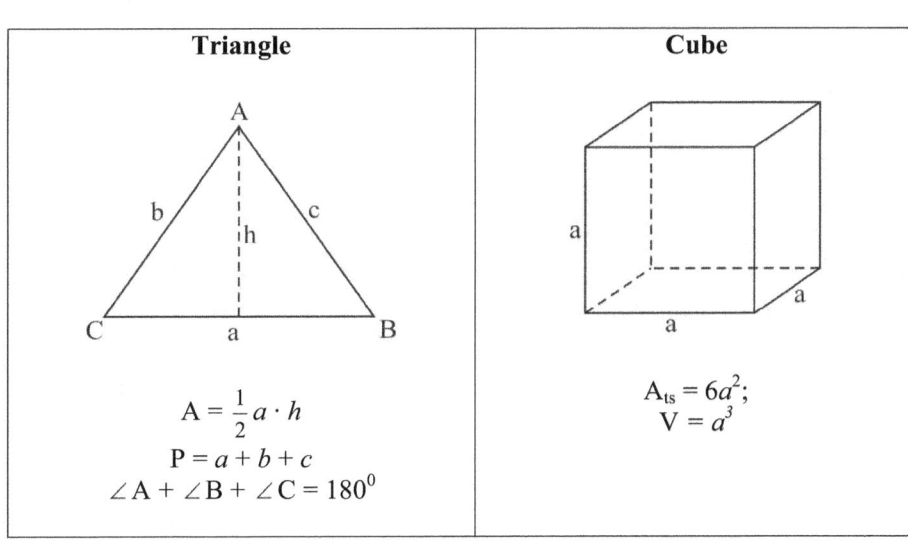

Triangle	Cube
$A = \dfrac{1}{2} a \cdot h$ $P = a + b + c$ $\angle A + \angle B + \angle C = 180^0$	$A_{ts} = 6a^2;$ $V = a^3$

Right Triangle	Triangular Prism
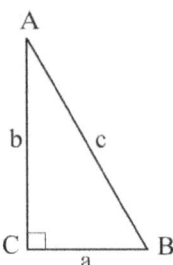	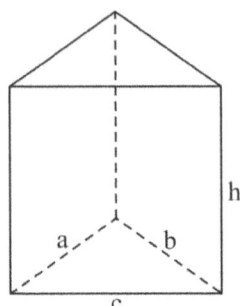
$A = \dfrac{1}{2} a \cdot b$ $P = a + b + c$ $\angle A + \angle B = 90^0$ $\angle C = 90^0$ $c^2 = a^2 + b^2$ $c = \sqrt{a^2 + b^2}$	$A_{ls} = (a + b + c) \cdot h = P_{base} \cdot h;$ $A_{base} = \dfrac{1}{2}\,(c \cdot h_b);$ h_b – height of the triangular base; $A_{ts} = A_{ls} + 2A_{base};$ $V = A_{base} \cdot h$
Rectangle	**Triangular Pyramid**
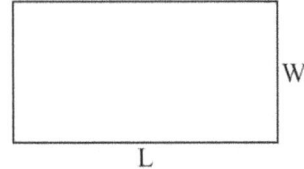	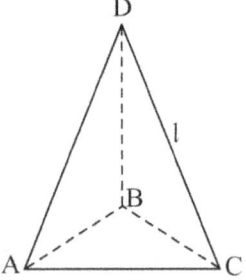
$A = l \cdot w$ $P = 2(l + w)$	$\mathbf{A_{ls}} = A_1 + A_2 + A_3,$ where A_1, A_2, A_3 – areas of lateral triangles (ADC, ADB, and BDC). $A_1 = \dfrac{1}{2} AC \cdot h_1;\ \ A_2 = \dfrac{1}{2} AB \cdot h_2;$ $A_3 = \dfrac{1}{2} BC \cdot h_3;$ h_1, h_2, h_3 – heights of lateral triangles ADC, ADB, and BDC; $A_{base} = \dfrac{1}{2}\,(AC \cdot h_b);$

	$A_{ts} = A_{ls} + A_{base};$ $V = \dfrac{1}{3} A_{base} \cdot h$
Square 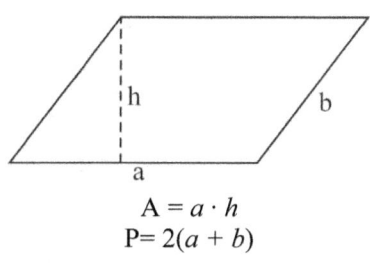 $A = a^2$ $P = 4a$	**Circle** 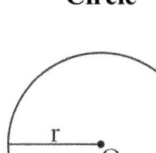 $\mathbf{c} = \pi d = 2\pi r;$ $A_{circle} = \pi r^2.$
Parallelogram $A = a \cdot h$ $P = 2(a + b)$	**Cylinder** 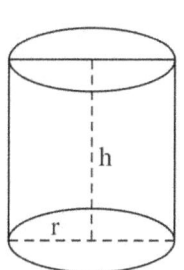 $A_{ls} = c \cdot h = 2\pi r h;$ $A_{base} = \pi r^2;$ $A_{ts} = A_{ls} + 2 A_{base};$ $V = A_{base} \cdot h = \pi r^2 h;$
Trapezoid 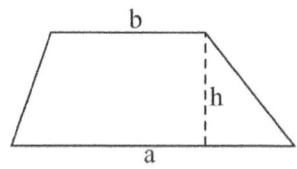 $A = \dfrac{1}{2}(a + b) \cdot h = \left(\dfrac{a+b}{2}\right) \cdot h$	**Cone** 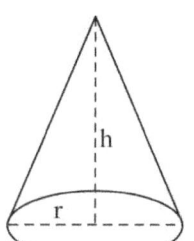 $A_{ls} = \dfrac{1}{2} l c = \pi r l;\; A_{base} = \pi r^2;$ $A_{ts} = A_{base} + A_{ls} = \pi r^2 + \pi r l =$

	$= \pi r (r + l);$ $V = \dfrac{1}{3} A_{base} \cdot h = \dfrac{1}{3} \pi r^2 h.$
Rectangular Parallelepiped 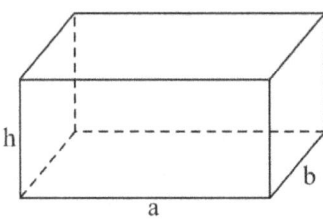 $A_{ls} = 2(a + b) \cdot h = P_{base} \cdot h;$ $A_{base} = a \cdot b;$ $A_{ts} = A_{ls} + 2A_{base}$ $V = a \cdot b \cdot h = l \cdot w \cdot h;$ where $a = l$, and $b = w$.	**Sphere** 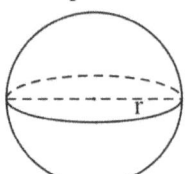 $A = 4A_{circle} = 4 \pi r^2;$ $V = \dfrac{4}{3} \pi r^3;$

Legend

a, b, c – sides
l = length
w = width
h = height
r = radius
V = volume
A = area

A_{ls} = area of the lateral surface
A_{base} = area of the base
A_{ts} = area of the total surface
P = perimeter

P_{base} = perimeter of the base
$c = 2 \pi r$ – length of circumference

ABOUT THE AUTHOR

J. Nyenetu Jarkloh is a Liberian-born graduate professional communications engineer, senior corporate manager and administrator, educator, author and publisher, corporate communications specialist, policy analyst, and a staunch advocate for human rights.

In Monrovia (Liberia), he attended the Amanda Caphart Elementary, Boatswain Junior High, Government Junior High, W. V. S. Tubman High, before enrolling at the University of Liberia (UL) in the College of Business. In the wake of Liberia's 1980 coup, based on excellent performance in his studies at the UL, he was granted a bilateral scholarship to pursue advanced studies in the former Soviet Union, where he earned a master's degree in radio engineering from the Odessa State Polytechnic University (Ukraine)

Upon return to his native country, Mr. Jarkloh had worked as a senior instructor in the Electronics Engineering Department at the W. V. S. Tubman Technical College (now Tubman Technical University) in Harper (Maryland County, Liberia), a part-time lecturer in mathematics (Numerical Analysis) at the University of Liberia, and also as Manager of Planning, Research & Development Department of the Liberia Telecommunications Corporation (L.T.C.) (now Liberia Telecommunications Authority). He also holds post-graduate professional certificates in telecommunications planning and management from the Satellite Transmission Systems Inc (Long Island, N.Y.), INTELSAT, and the United States Telecommunications Training Institute (Washington, D. C., U.S.A).

As a Fellow of the International Telecommunications Union (ITU), Mr. Jarkloh had actively participated in the ITU's GMDSS project, which called for elaboration of national master plans from English-speaking African Countries for the Development of Maritime Radiocommunications services. He elaborated his country's "National Master Plan for the Development of Maritime Radiocommunication Services in Liberia", which he defended during the final workshop of said project at the International Center for Theoretical Physics (ICTP) in Trieste (Italy). Due to the Liberian Civil Crisis, however, he was evacuated. He presently resides with his family in Ukraine.

www.ingramcontent.com/pod-product-compliance
Lightning Source LLC
Chambersburg PA
CBHW051450170526
45166CB00001B/190